W9-DJO-338

A Special Issue of
*The European Journal
of Cognitive Psychology*

Bridging Cognitive Science and Education: Learning, Memory, and Metacognition

Guest Editors

Lisa K. Son
Barnard College, USA

and

André Vandierendonck
Ghent University, Belgium

Psychology Press
Taylor & Francis Group

HOVE AND NEW YORK

First published 2007 by Psychology Press
27 Church Road, Hove, East Sussex, BN3 2FA

www.psypress.com

Simultaneously published in the USA and Canada
by Psychology Press
270 Madison Avenue, New York, NY 10016

Psychology Press is part of the Taylor & Francis Group, an Informa business

© 2007 Psychology Press

British Library Cataloguing in Publication Data
A catalogue record for this book is available from the British Library

ISBN 13: 978-1-84169-835-9
ISSN 0954-1446

Cover design by Anń Design, Tara, Co. Meath, Ireland
Typeset in Ireland by Datapage International, Dublin
Printed in the UK by Hobbs the Printers Ltd, Totton, Hampshire
Bound in the UK by TJ International Ltd, Padstow, Cornwall

The publication has been produced with paper manufactured to strict
environmental standards and with pulp derived from sustainable forests.

Contents*

(*continued overleaf*)

---

* This book is also a special issue of *The European Journal of Cognitive Psychology*, and comprises pp. 481–768 of Volume 19 (2007). The page numbers are taken from the journal and so begin with p. 481.

EUROPEAN JOURNAL OF COGNITIVE PSYCHOLOGY
2007, 19 (4/5), 481–493

Introduction: A metacognition bridge

Lisa K. Son

Barnard College, New York, NY, USA

We are told never to cross a bridge until we come to it, but this world is owned by men who have "crossed bridges" in their imagination far ahead of the crowd. (Anon.)

John Dewey, a well-known education theorist and pragmatist, said the following: "Without insight into the psychological structure and activities of the individual, the educative process will, therefore, be haphazard and arbitrary" (1897, p. 77). For a long time, the two disciplines, *cognitive science* and *education*, have worked hard to discover effective principles of learning with the goal of improving educational achievement. And although each has made significant advances, there has been a gap between the two disciplines, a gap that remains a reality today. In this introduction, I provide a brief synopsis of the historical paths towards the understanding of learning in both the cognitive field and educational realm, with the aim to show that there is a need, now more than ever, to throw a bridge between the two fields. In this special issue, *Bridging Cognitive Science and Education: Learning, Memory, and Metacognition*, we have organised and integrated some of the recent cognitive data that can provide some new insights on the complexities of learning in the classroom.

The question of how people learn, be it inside or outside the classroom, has been a major theme in cognitive science. Although the cognitive revolution is thought to have arrived in the mid-1900s, *cognitive science,*

Correspondence should be addressed to Lisa K. Son, Department of Psychology, Barnard College, 3009 Broadway, New York, NY 10027, USA. E-mail: lson@barnard.edu

I thank Nate Kornell and Stephen Peverly, as well as my co-editor on this special issue, André Vandierendonck, for comments and suggestions on earlier drafts of this introduction. I am also grateful to the Spencer Foundation, who inspired the organisation of this special issue, and to Barnard College, who hosted and funded a special conference called *Metacognition: Theory and Application*, after which the idea for the special issue evolved. In the writing of this article, the following websites were consulted: and http://www.bls.org/cfml/l3tmpl_history.cfm; and http://www.infed.org/thinkers/et-froeb.htm

DOI: 10.1080/09541440701352317

which is defined by empirical investigation, was instigated much earlier. Francis Bacon, who lived from 1561 to 1626, was the first to develop the practice of "observe and experiment" as the method of scientific inquiry. Later, in 1879, Wilhelm Wundt set up the first psychology laboratory, giving him the title "founder of experimental psychology". And soon thereafter, Hermann Ebbinghaus (1885/1962) performed the first empirical studies on remembering and forgetting, using himself as his own guinea pig. Around the same time, and continuing into the twentieth century, work of the behaviourists, as well as a list of more contemporary pioneers of the cognitive revolution—Hebb (1949) with his insight into the neural mechanisms of brain activity, Miller with his discovery of our "magic" short-term capacity (1956), and Chomsky (1959) with his influential theories of linguistics and mental states, to name but a few—paved the way for decades of continued investigations into how the process of learning functions (and malfunctions).

Throughout this extensive history of empirical investigation, as summarised in the quick and easy synopsis above, terms such as *encoding, short-* and *long-term storage, transfer appropriate processing, rehearsal,* and many others, became household names to cognitive scientists. Years of data collection in the laboratories made researchers aware of what types of strategy could potentially improve learning, as well as those types that could hurt learning. For example, it was recognised that we should *space,* and not *mass* our learning sessions (e.g., Melton, 1970); that we should actively *generate,* and not passively *read* the items that we wanted to learn (e.g., Slamecka & Graf, 1978); if possible, we should *cluster* or *chunk* a list of items that needed to be remembered (e.g., Simon, 1974), and we should exercise interactive *imagery* rather than *rote* maintenance for enhanced learning (e.g., Craik & Lockhart, 1972). And finally, a new subfield emerged, one where researchers focused on a framework of "metacognition", which included *monitoring* and *control* of one's own learning. Still, research was largely limited to the laboratories, and it was not until only recently that cognitive researchers focused on learning inside the classrooms.

Within the education realm, of course, understanding and improving learning has been a high priority. In America even as early as the 1600s, for instance, everyone was expected to learn to read, if at the least for religious reasons, and it was not long before an organised place where children would go to learn evolved—a *school*. In 1635, the Boston Latin School, considered to be the first school in the US, opened for boys. The next few decades saw a nationwide expansion of schools. In most of these early "one-room" schoolhouses, students were encouraged to behave in a way that was composed and respectful, especially to the teachers. In the classroom, teachers were meant to be in charge. It was still a century too early for the idea that *student-centred* learning might be more effective than learning that was *teacher centred*.

With the expansion of organised learning facilities came the development of *educational psychology*, a branch aimed to improve learning and, in particular, school performance. Like the goals of cognitive psychologists, educational psychologists also aimed to better understand how learning occurred and how the process could be improved. Friedrich Frobel, who lived from 1782 to 1852, was one of the first educational psychologists, and the first to believe that children should not be reserved, that *activity* was key in the learning of a child. These ideas went directly against the practices of the early schoolhouses, where students had been expected to be inactive. Soon thereafter, John Dewey published "My Pedagogic Creed" (1897), and became a major critic and reformer of the earlier school philosophies throughout the early 1900s. In his pedagogic creed, Dewey wrote that the teacher's task was not to impose certain ideas, but instead to merely stimulate and guide the learner. Already, the notion was rising that students, not teachers, should be the centre of learning.

It was also apparent that success in the classroom meant that students had to perform at some criterion level on a variety of tests. As a result, educators became concerned with evaluation and assessment in the schools, and hoped to be able to diagnose all students, including those who might have had particular learning problems. In 1905, Alfred Binet developed the first *IQ test*, a testing method that became extremely influential in education (although years later, in the 1980s, the meaningfulness of the IQ test was challenged by Howard Gardner's *theory of multiple intelligences*; 1983). Standardised testing became a common practice, and by the 1970s and 1980s, even primary grade children were tested. Although many were opposed to such educational policies, there was a clear understanding that school learning and performance would have a dramatic impact on a child's life, and thus, it was vital that we develop the type of classroom atmosphere that would be most amenable to successful learning.

Throughout the twentieth century, educational psychologists, in collaboration with developmental psychologists and educators, investigated the types of practices that had the potential of improving learning in a classroom setting. And they, like cognitive researchers, also had their own set of household names and methods. Many of these methods were described as *metacognitive* strategies, although again, the definition of "metacognition" used by educators was different (and distinct) than that used by cognitive researchers described above. Here, techniques such as *note-taking*, *outlining*, and *summarising* became important issues to examine (e.g., Anderson & Armbruster, 1984). Other manipulations such as *clarifying* and *predicting* also were thought to be beneficial metacognitive devices (e.g., Brown & Campione, 1990). The relationships between members of an entire classroom, rather than the processes of simply one individual, were also observed as a metacognitive activity. For instance, *reciprocal teaching*, or the method in which teachers and students have a back-and-forth dialogue about a particular text, was found to facilitate learning (e.g., Palincsar & Brown, 1984). Similarly, *cooperative learning*, which encouraged students of different levels to work in small face-to-face groups, was acknowledged both in the US and internationally (e.g., Cohen, 1994). Finally, the effect of *feedback*, defined as teacher responses that affect student learning, was also a significant issue (e.g., Brophy & Good, 1986).

Both cognitive researchers and educators had accumulated a sizeable amount of information. However, many feel that the two fields, along with their independent findings, remain separate for the most part (e.g., Kuhn & Dean, 2004). In order to further knowledge on the issue of learning and performance, there is a strong need for communication between the two fields, and this is precisely the aim of this special issue.

CROSSING BRIDGES

Although cognitive research and education were distinct, there are a number of individuals who might be credited for designing and crossing the earliest bridges between the two fields. And interestingly, although the term "metacognition" had been defined already in different ways by the separate fields, it was also via another general description of metacognitive processes—the ability for an individual to reflect on his or her own thinking (Flavell, 1979)—that early bridges were crossed.

Several well-known individuals and their theories paved the way for the emergence of a new metacognitive field. These individuals were tremendously influential for both cognitive psychologists (and the devising of laboratory experiments), as well as for educators (and the structure of learning in the classroom). Jean Piaget was perhaps the most influential, particularly through the 1970s and 1980s. Piaget (1972) is best known for classifying the stages of cognitive development, starting with infancy, and continuing through to adolescence. The idea behind his stages was that a child learned as a result of reorganising and reflecting on the knowledge gained at each level, including knowledge of errors. Piaget's theories had a long-lasting effect on educational policy such that learning became more student centred, but also affected cognitive science in that age-comparison studies could provide valuable methods by which to capture the various step-by-step mechanisms that are required for learning and self-reflection.

Perhaps influenced by Piaget, another recognised "bridge builder" was Lev Vygotsky, made most famous with his contributions on the *zone of proximal development*. In his book *Mind and Society: The development of higher mental processes* (1930/1978), Vygotsky focused on student-centred learning, but in collaboration with guidance from the adult who would act as a "scaffold". He believed that there was a "zone", defined by a difference between what a child could learn on his own and what the child could learn with meaningful help from an adult. Both interested in student-centred learning, Piaget and Vygotsky were key figures in the emergence of how children think about their own learning.

Although Piaget and Vygotsky had laid the groundwork, the study of metacognition in both the laboratories and the classrooms became front and centre largely because of another researcher, John Flavell. Flavell was a psychologist who had long been interested in Piaget's work, and in the understanding of a child's mental states. In fact, his book *The Developmental Psychology of Jean Piaget* (1963) was the first most comprehensive evaluation of Piaget's work in English. In the 1970s and 1980s, Flavell focused his research on evaluating children's role-taking abilities, or the abilities that children had to understand what another person knew. He was also interested in the development of memory in children, and found that children needed to understand the concept of memory before they could actually develop skills for applying or improving memory processes. This higher knowledge, of understanding memory, he officially called *metamemory* (1971).

Since Flavell's introduction of the term, metamemory and metacognitive research continued to expand. Answers to questions that had been posed centuries earlier about how to improve learning and memory are being sought, and more importantly, are sought in forms that are realistic and effective for learning in the classrooms, not just in the laboratories. Researchers and educators together are now motivated to discover the most optimal practices for learning, memory, and metacognition.

This special issue shows that a growing number of researchers are interested in crossing the "metacognition bridge", and are devoting a good part of their investigations to understanding strategies that enhance learning and memory, and the awareness of those strategies, particularly in real-world educational situations. The main purpose for this collection of papers is to have, in one place, an encyclopaedia of information that will provide researchers and educators with new findings about how people learn and remember information. In organising the issue, I have categorised the papers into two broad categories that aim to answer a few general questions. The first is "What types of strategy improve learning?" and is addressed by the first five papers. And the second is "Are people aware of such strategies?", addressed by the next seven papers in this issue. This set of seven papers will also relate to issues such as what kinds of people are aware as well as the relation between awareness and learning. The final paper in the series suggests that when an appropriate "metacognitive" scaffold is used—one that incorporates a variety of effective cognitive strategies—learning in the student will improve enormously. Below, I summarise briefly the topic in each paper.

The first paper uses a method of research that bridges directly cognitive research and the classroom. In their paper, the authors McDaniel, Anderson, Derbish, and Morrisette investigate the *testing effect* in an everyday college course. Typically, people think of testing as a device that merely measures performance (e.g., Kornell & Bjork, in press; Kornell & Son, 2006), but recently, people have shown that testing has benefits for the process of learning too (e.g., McDaniel & Fisher, 1991). In this research, students took weekly quizzes made up of multiple choice (MC) tests or short answer (SA) tests. Their results showed that quizzing (as compared to those not quizzed at all) improved performance, and that short answer quizzes improved performance the most, especially when accompanied by corrective feedback.

In the next paper, Butler and Roediger examine the testing effect and how it affects performance over the long term. Participants were given initial exposure to a classroom lecture via a video. Then, participants either took an MC test or an SA test, or simply studied a summary of the lecture. A final SA test was administered after 1 month. The results indicated that an initial SA test produced final test performance that was superior to that resulting from either an MC test or additional study. Kang, McDermott, and Roediger also study the effects of testing in the next paper, and in this case, had participants learning papers on their own. After reading the papers, participants received an MC test or an SA test, or a list of statements to read, or a filler task. Their question was: Do MC or SA tests differentially improve performance on later tests, and if so does the type of final test matter? The results showed that when feedback was available, there was an advantage for the SA test which yielded the most benefits on later tests, regardless of the type of final test.

An important issue in the investigation of learning is feedback. How much feedback is effective for learning and does feedback matter? Rawson and Dunlosky focus on this question in their paper, where they test people's learning of text passages. They also asked whether people's metacognitive knowledge improved when feedback was given versus when it was not given. In their experiment, college students read passages, each of which included key terms. After reading each text, each key term was presented, and participants attempted to recall the corresponding definition. Then, they were given the opportunity to "score" their own answers. Some were given feedback before scoring their answers—that is, they were shown the correct answer, in the "standard" condition; others were not able to see the correct answer—the "no standard" condition. The results showed that both recall performance and metacognitive accuracy improved when participants were able to compare their own answers to the correct answer during the "self-scoring" stage. And, although all participants were overconfident in the correctness of their answer, the "standard" participants were less so.

Carroll, Campbell-Ratcliffe, Murnane, and Perfect then ask if a higher level of expertise could protect against forgetting. In order to test forgetting, they used the *retrieval-induced forgetting* paradigm (e.g., Anderson & Spellman, 1995). Retrieval-induced forgetting occurs when memory for items that are not rehearsed are remembered less well than items that were rehearsed in that same category as well as items from other non-rehearsed categories. For instance, if people rehearsed *banana* and *plum*, then memory for *pineapple* and *melon* would be worse than memory for *banana* and *plum*, as well as for *chair* and *table*. This "forgetting" of *pineapple* and *melon* has been thought to be a type of unconscious forgetting. The question the authors addressed here was: Are people aware of retrieval-induced forgetting? If so, can something be done to compensate for that forgetting? For instance, might a student spend extra time trying to study the *connections* between items within one particular category? In one experiment, comparing experts versus novices (who would have less awareness of connections between items in one category), they found that, indeed, experts were somewhat protected from retrieval-induced forgetting, especially when there was not enough time during study to form new connections. In addition, the expert participants were more aware of what they would remember and not remember. That is, the experts thought there would be less inhibition than the novices did. In the second experiment, they follow up similar investigations on the influences of test format.

A few of the early papers, while investigating the types of strategies that enhance learning, have also examined how aware or attentive people are of what they know and don't know. Groups have different levels of attention, for a variety of reasons. For instance, some have attentional disorders, others have high levels of anxiety. And it is certainly likely that students in school will have different levels of attention or anxiety when it comes to testing and performance. One question of interest then is: How will these differences manifest in learning and metacognitive processes? Moreover, is there a developmental component to the metacognitive process? For instance, do younger children have more difficulty with knowing what they know and do not know? And for that matter, are there differences in monitoring abilities between people of different levels of experience or expertise? The following set of papers all address these issues.

In their paper, Ballesteros, Reales, and Garcia examine recognition and priming of attended and unattended stimuli at encoding in two different ages (second and fifth grades) of both normal children and those with attentional deficits (AD). In the recognition phase, children were asked to name an object that was presented in a particular colour and overlapped with another object in a different colour. The named objects were the attended items while the ignored overlapping objects were "unattended". In the test phase, they performed a picture fragment completion task (PFCT) that included both attended and unattended objects. For instance, they were shown gradually more filled in pictures until they were able to identify the item. Results showed while the AD children suffer from expressing their unattended or "implicit" memory in general, they show similar patterns of attention development as normal controls.

Meneghetti, De Beni, and Cornoldi also examine adolescents, and their use of effective and less effective strategies. From a sample of 354 students, aged 11 to 15, two groups of students were selected based on their performance on a reading comprehension task. The goal of the study was to compare knowledge in the two groups, anticipating that the high scoring reading comprehension group would show coordination between knowledge and use of strategies, the low scoring group a greater discrepancy between the two ratings. And indeed, this is what they found. Such findings implicate the importance of metacognition in strategy use and consequent academic performance.

Miesner and Maki follow up with the question of how anxiety influences metacognitive monitoring. Many feel that anxiety is a negative emotion. Feeling too nervous or anxious might distract from your attention or your learning. Meisner and Maki investigate whether anxiety might help or hurt your monitoring performance as well. For instance, if you are anxious about your knowledge, does that mean that you are a worse or better predictor of what you believe you know? After rating people using the Test Anxiety Scale, people were given texts to read for a later test. Before taking the test, they were also asked to judge how well they thought they would do on the test as a measure of monitoring accuracy. Results showed that high anxiety did hurt performance, as well as decrease confidence. Interestingly, people high in anxiety were more accurate on metacomprehension, particularly on MC tests.

De Bruin, Rikers, and Schmidt also test monitoring accuracy in learners at different expertise levels—experienced and inexperienced chess players. In their study, all participants were presented with a variety of chess rules, and had a learning phase where they saw a particular move. Then they made judgements about how well they would be able to predict such moves in the future. They were given a final test in the format of a new game. Results showed that experienced chess players not only learned faster and performed better, but also had higher monitoring accuracy.

Kelemen, Winningham, and Weaver split their participants as a function of SAT scores, and compare monitoring ability. Subjects were given foreign vocabulary pairs to learn and then make judgements about how well they thought they would remember the pairs later on the final test. Then they were given the final test. Those with higher SAT scores had higher performance on the test, and were less overconfident. In addition, training across trials only benefited the students who had higher SAT scores. In a series of additional experiments, they investigate how monitoring accuracy might be improved. For instance, they find that practice can improve monitoring accuracy, without any explicit instructions or added feedback, and that mere exposure to study and recall procedures, without having to make metacognitive judgements, also improved monitoring accuracy.

Higham and Arnold investigate people's awareness of their knowledge during MC tests using a *signal-detection* approach. On MC tests, learners need to know which answer is correct and which answers are incorrect. If some particular option has reached some criterion of confidence as correct, then that answer is chosen as the correct one. If none of the choices reach criterion, then one could choose to withhold a response and not answer the question. Using signal detection, the authors were able to see who has a conservative criterion and who has a liberal criterion—which could be used a metacognitive monitoring measure. Their results showed that in general, people were underconfident—that is they withheld too often—and that there was no improvement across time. More importantly, the lowest-scoring groups of students did not respond in a manner that was consistent with their actual performance. Those who monitored well performed better than those who monitored more poorly. These results, in combination to the previous findings in this special issue, suggest that awareness of one's own learning, or the enhancement of metacognitive accuracy, can improve learning itself.

In the final paper of this special issue, my co-authors Metcalfe and Kornell, and I, show that when a computer acts as an effective "metacognitive" scaffold for a student by introducing practices that use cognitive principles to enhance learning, learning in the student will improve enormously. In the research, students in both middle school and college either studied on their own, or on a computer program in which spacing strategies, generation strategies, and contextual processing strategies were enforced. And one of the most valuable aspects of the study was that the procedure took place over a long term, across a total of 7 weeks. The final results show a huge improvement on the final test following computer study over self-study. One could interpret these results as showing that individuals are indeed, bad at learning. On the other hand, one could rest assured that even when students use suboptimal strategies, and even when they do not have accurate metacognitive knowledge, we can still rely on educational implementations (and teachers) that are knowledgeable about effective learning strategies and can appropriately guide students during learning. These data, and the information gained in all of the papers in this issue, show that the bridge between cognitive science and education is crucial for an individual's success in school.

TRAFFIC ON THE BRIDGE

Historically, both cognitive researchers and educators have invested huge amounts of energy and time to our understanding of learning. Strategies that enhance learning both in the laboratories and in the classrooms have been tested and applied in many ways, and continue to expand. And metacognition, what I describe here as an early bridge between the two fields, has grown to be a theme that, at the moment, seems to have numerous empirical answers for the improvement of real-world learning. Perhaps a question that remains is this: How long will researchers investigate real-world learning paradigms, and similarly, how long will educators be interested in the research? In other words, can we expect a lot of traffic across this bridge?

I offer this question for several reasons. First, there has already been a long history of a largely divided path of research—one for cognitive researchers and another for educators. And it is likely that there are several causes of the creation of this divide. One is that there are still few opportunities for researchers and educators to speak to each other. Over the past semester, I was able to visit a local private school, and survey anonymously a group of 20 teachers (who teach a variety of subjects and a variety of grades in elementary and middle school). One of the questions on that survey was: "Have you ever spoken before with any cognitive researchers (besides me) who conduct research on learning strategies?" Out of 20, only 4 said that they have spoken about some research findings. On the other hand, all 20 answered "Yes" when asked: "Do you think that your teaching strategies would improve if you were made aware of various cognitive strategies?" and "Do you think that it is important to understand the psychological mechanisms that underlie the different learning strategies?"

Also on that survey, the teachers were asked if they knew of such practices such as the *generation effect*, the *spacing effect*, and *levels of processing*. Out of all of their answers, only 15%, 20%, and 50%, respectively, knew that these methods had been found to enhance learning in the laboratories. After describing each of these practices in some detail, I asked the teachers whether they ever used these tactics in the classroom, and how many of their students would use these strategies on their own, in their opinion. On average, about 80% said they used these strategies in the classroom (each of the above strategies were used at around the same frequency), but felt that only about 20% of their students would use such strategies on their own. And all of the teachers agreed that learning could be much improved at the level of the individual learner, rather than in the classroom.

Another reason for why the two fields have stayed fairly divided is that educators have long felt that the data collected in the laboratories have not been applicable to their own classrooms and the subjects that they teach. For instance, many of the teachers expressed that they are not interested in having their students learn a list of words. When asked in the survey "Do you think that connecting with cognitive researchers would be helpful for the teaching strategies in your subject area?" one teacher said "not at all", two teachers said "a little", four said "not sure", ten said "probably a lot", and only three said "definitely". Given these answers, it seems that there would be a varying range of how valuable this bridge of data and the classroom would be.

A few teachers expressed to me that researchers should realise that the types of material that need to be learned in school are often complex, especially to a child. When asked on the survey "Do you think that cognitive researchers could learn from speaking with you and your teaching strategies?" seven teachers said "definitely", eight said "probably a lot", but five said that they were "not sure". (None of the teachers chose "not at all" or "a little".)

Surprisingly, 19 out of the 20 teachers surveyed had read some number of scientific papers that addressed learning prior to or during their teaching careers. However, a few said that cognitive scientists write in a way that was "boring" or "cryptic". And the larger problem was that the learning scenarios were not realistic, and the materials were not typical of what a student in their class would have to learn. One even said that "it's obvious that cognitive researchers who collect data in the lab have very little knowledge of what goes on in the classroom". When the teachers were asked "Do you think that reading scientific papers would be helpful for your teaching strategies?" there was a range of answers: One teacher said "not at all", two said "a little", eleven said "not sure", five said "probably a lot", and five said "definitely".

Even though there may be some uncertainty about whether traffic on this bridge between cognitive science and education will continue to increase, I believe that it will. Fortunately, many cognitive researchers, including those who have contributed to this special issue, have put effort into asking teachers and educators questions about learning and materials in their classrooms. And using that knowledge, they have designed new and applicable experiments. I also feel that the main reason that teachers have found some scientific papers to be "boring" or "cryptic" is simply because they have not been successfully applicable to classroom learning. In this special issue, I hope this will not be the case. The researchers have, I feel, successfully tested the various means of learning in highly realistic and valuable ways.

Since the beginning of cognitive research and the beginning of education, even as separate fields, we have not done too badly. Given what Dewey might have seen as a "haphazard and arbitrary" educational process, children and adults have been able to perform fairly well in the hundreds of exams that they take in their lifetimes. But he is correct to believe that improved communication between the two fields can only advance the state of our knowledge. The contributors to this special issue have made valiant efforts in testing learning as it pans out in the real world, and, through this special issue, we hope to disperse that information with the goal of continued links between cognitive scientists and educators. In short, I am looking forward to an unbreakable, and congested, bridge.

REFERENCES

Anderson, M. C., & Spellman, B. A. (1995). On the status of inhibitory mechanisms in cognition: Memory retrieval as a model case. *Psychological Review, 102*, 68–100.

Anderson, T. H., & Armbruster, B. B. (1984). Studying. In P. D. Pearson (Ed.), *Handbook of reading research* (pp. 319–352). New York: Longman.

Binet, A. (1905). New methods for the diagnosis of the intellectual level of subnormals. *L' Année Psychologique, 12*, 191–244.

Brophy, J. E., & Good, T. L. (1986). Teacher behavior and student achievement. In M. C. Wittrock (Ed.), *Handbook of research on teaching* (3rd ed., pp. 328–377). New York: Macmillan.

Brown, A. L., & Campione, J. C. (1990). Communities of learning and thinking, or a context by any other name. In D. Kuhn (Ed.), *Contributions to human development: Vol. 21. Developmental perspectives on teaching and learning thinking skills* (pp. 108–126). Basel, Switzerland: Karger.

Chomsky, N. (1959). A review of B. F. Skinner's verbal behavior. *Language, 35*, 26–58.

Cohen, E. G. (1994). *Designing groupwork: Strategies for the heterogeneous classroom.* New York: Teachers College Press.

Craik, F., & Lockhart, R. (1972). Levels of processing: A framework for memory research. *Journal of Verbal Learning and Verbal Behavior, 11*, 671–684.

Dewey, J. (1897). My pedagogic creed. *School Journal, 44*, 77–80.

Ebbinghaus, H. (1962). *Memory: A contribution to experimental psychology.* New York: Dover. (Original work published 1885)

Flavell, J. H. (1963). *The developmental psychology of Jean Piaget.* New York: Van Nostrand Reinhold.

Flavell, J. H. (1971). First discussant's comments: What is memory development the development of? *Human Development, 14*, 272–278.

Flavell, J. H. (1979). Metacognition and cognitive monitoring: A new area of cognitive-developmental inquiry. *The American Psychologist, 34*, 906–911.

Gardner, H. (1983). *Frames of mind: The theory of multiple intelligences.* New York: Basic.

Hebb, D. O. (1949). *The organization of behavior.* New York: Wiley.

Kornell, N., & Bjork, R. A. (in press). The promise and perils of self-regulated study. *Psychonomic Bulletin and Review.*

Kornell, N., & Son, L. K (2006). *Self-testing: A metacognitive disconnect between memory monitoring and study choice.* Poster presented at the 47th annual meeting of the Psychonomic Society, Houston, TX.

Kuhn, D., & Dean, D., Jr. (2004). Metacognition: A bridge between cognitive psychology and educational practice. *Theory into Practice, 43*, 268–273.

McDaniel, M. A., & Fisher, R. P. (1991). Tests and test feedback as learning sources. *Contemporary Educational Psychology, 16*, 192–201.

Melton, A. W. (1970). The situation with respect to the spacing of repetitions and memory. *Journal of Verbal Learning and Verbal Behavior, 9*, 596–606.

Miller, G. A. (1956). The magical number seven, plus or minus two: Some limits on our capacity for processing information. *Psychological Review, 63*, 81–97.

Palincsar, A. S., & Brown, A. L. (1984). Reciprocal teaching of comprehension-fostering and comprehension-monitoring activities. *Cognition and Instruction, 1*, 117–175.

Piaget, J. (1972). Intellectual evolution from adolescence to adulthood. *Human Development, 15*, 1–12.

Simon, H. (1974). How big is a chunk? *Science, 183*, 482–488.

Slamecka, N. J., & Graf, R. (1978). The generation effect: Delineation of a phenomenon. *Journal of Experimental Psychology: Human Learning and Memory, 4*, 592–604.

Vygotsky, L. S. (1978). *Mind and society: The development of higher mental processes.* Cambridge, MA: Harvard University Press. (Original work published 1930)

EUROPEAN JOURNAL OF COGNITIVE PSYCHOLOGY
2007, 19 (4/5), 494–513

Testing the testing effect in the classroom

Mark A. McDaniel

Washington University in St Louis, St Louis, MO, USA

Janis L. Anderson

Harvard Medical School and Brigham & Women's Hospital, Boston, MA, USA

Mary H. Derbish

Washington University in St Louis, St Louis, MO, USA

Nova Morrisette

University of New Mexico, Albuquerque, NM, USA

Laboratory studies show that taking a test on studied material promotes subsequent learning and retention of that material on a final test (termed the testing effect). Educational research has virtually ignored testing as a technique to improve classroom learning. We investigated the testing effect in a college course. Students took weekly quizzes followed by multiple choice criterial tests (unit tests and a cumulative final). Weekly quizzes included multiple choice or short answer questions, after which feedback was provided. As an exposure control, in some weeks students were presented target material for additional reading. Quizzing, but not additional reading, improved performance on the criterial tests relative to material not targeted by quizzes. Further, short answer quizzes produced more robust benefits than multiple choice quizzes. This pattern converges with laboratory findings showing that recall tests are more beneficial than recognition tests for subsequent memory performance. We conclude that in the classroom testing can be used to promote learning, not just to evaluate learning.

Correspondence should be addressed to Mark A. McDaniel, Department of Psychology, Campus Box 1125, Washington University in St Louis, St Louis, MO 63130, USA. E-mail: mmcdanie@artsci.wustl.edu

This research was supported by Institute of Educational Sciences Grant R305H030339. Mary Derbish's participation was supported by a Collaborative Activity Grant from the James S. McDonnell Foundation. This experiment was presented in part at the 46th Annual Meeting of the Psychonomic Society, Toronto, Canada, November 2005 and at the Annual Meeting of the American Educational Research Association, San Francisco, California, April 2006. We thank Roddy Roediger and Chuck Weaver for helpful comments on an earlier version of this paper and Austin McDaniel, Jesse McDaniel, and Rebecca Roediger for assistance with aspects of the data scoring.

In educational practice, as well as in the educational research literature, testing has been primarily considered an evaluative instrument. However, many researchers who study memory have considered testing from the perspective of its mnemonic benefits. Experimental reports have repeatedly demonstrated that taking a test on studied material promotes subsequent learning and retention of that material on a final test (e.g., Bartlett, 1977; Darley & Murdock, 1971; Hanawalt & Tarr, 1961; Hogan & Kintsch, 1971; Masson & McDaniel, 1981; McDaniel, Kowitz, & Dunay, 1989; McDaniel & Masson, 1985; Whitten & Bjork, 1977). For purposes of exposition we will refer to the memory gains produced by intervening tests as the *testing effect*. Experimental memory research has established that the testing effect is robust across materials and types of tests. Testing effects are observed with word lists (Hogan & Kintsch, 1971; McDaniel & Masson, 1985), paired associate lists (Allen, Mahler, & Estes, 1969; Carrier & Pashler, 1992), pictures (Wheeler & Roediger, 1992), and prose material (Glover, 1989; Roediger & Karpicke, 2006b). Testing effects surface when the intervening tests are different from the final tests: intervening recall tests improve subsequent recognition (Glover, 1989; Lockhart, 1975; Wenger, Thompson & Bartling, 1980) and intervening recognition tests improve subsequent recall (Runquist, 1983). Finally, taking a test is almost always a more potent learning device than additional study of the target material (see Carrier & Pashler, 1992, for recent experimental tests, and Roediger & Karpicke, 2006a, for a review).

Despite this impressive body of evidence, the implications of the testing effect literature for educational practice have been virtually ignored by the educational community and educational research. Echoing this observation, an educationally relevant study on the testing effect was entitled "The 'testing' phenomenon: Not gone but nearly forgotten" (Glover, 1989). Yet, paralleling the basic memory findings, the few studies in the educational literature that have examined the testing effect have found positive benefits of intervening tests on final test performance (Glover, 1989; Spitzer, 1939). Despite these findings, current texts on learning and instruction fail to mention the possible merits of using tests to potentiate learning and retention (e.g., Mayer, 2003; see also Baine, 1986). This omission may be warranted because even the studies appearing in the educational psychology literature have not demonstrated the benefits of testing on material being taught and learned in the classroom.

To fill this critical gap, the purpose of the present research was to experimentally examine the testing effect for content presented throughout the semester in a college course. We were interested in several overarching issues. First, would positive testing effects emerge in the context of a standard course? Testing in a course diverges in important ways from the implementation of testing in laboratory studies. In a class, there is presumably great variability across students in the amount of studying of the target material and in the interval between study and the intervening testing (testing prior to the criterial tests). In contrast, both of these variables are carefully controlled or manipulated in laboratory studies. Further, the delays between intervening test (i.e., quizzes) and final criterial tests in the classroom can be on the order of days, weeks, or even months. In the laboratory long retention intervals are typically 1 or 2 days (Carrier & Pashler, 1992; Hogan & Kintsch, 1971; Masson & McDaniel, 1981; McDaniel & Masson, 1985); more often these intervals have been on the order of minutes or hours (e.g., Bartlett, 1977). In only very few experiments have benefits of testing in laboratory studies been examined at 1-week or longer intervals (see Roediger & Karpicke, 2006b, and Wheeler, Ewers, & Buonanno, 2003, for 1-week delay studies, and Butler & Roediger, 2007, this issue, for a 1-month delay). Because of these differences between the class environment and the parameters of the laboratory research, it is not certain that the testing effect will generalise to the more variable environment of an actual class.

For the present study, we identified several key issues from the existing literature that could be particularly important for motivating and implementing testing as a learning tool in the classroom. We next briefly review these issues and related findings in developing the rationale for the design of the current experiment.

EXTENDING TESTING EFFECTS TO THE CLASSROOM

One central issue raised by the testing effect findings is the extent to which the repeated exposure of the material stimulated by tests plays a role in the positive impact of intervening tests on final test performance. Some prominent studies in the educational literature do not clarify this issue because conditions were not included that receive extra study instead of intervening tests (e.g., Glover, 1989; Spitzer, 1939). Findings from the basic experimental literature do suggest that testing produces learning/retention advantages beyond that enjoyed from repeated study (provided that the retention intervals between intervening and final testing are greater than several minutes; cf. Roediger & Karpicke, 2006b; Wheeler et al., 2003). For instance, immediate retrieval of once-studied target items benefits performance on subsequent tests more so than does another study presentation of the target material (Hanawalt & Tarr, 1961; Hogan & Kintsch, 1971; McDaniel & Masson, 1985).

To assess the degree to which testing effects in the classroom (if found) reflect mnemonic processes that are more than just additional exposure of the content, in the current experiment we included an exposure-only condition in which the target facts were presented for reading. In this read only condition, participants were presented with the same information that was quizzed in other conditions. In addition, a control was implemented in which some facts in the course were neither presented for additional reading nor testing. Based on the literature cited, we expected testing effects to emerge (quizzes with feedback would produce better performance on final tests than not tested/read facts), and we expected that testing (quizzing with feedback) would be superior to the read only presentation in terms of boosting performance on final testing.

Another issue addressed in the basic memory literature is the relative benefit of cued recall tests over recognition (e.g., multiple choice) tests. Studies with simple laboratory materials (word or paired associate lists) have found that retrieval through recall benefits subsequent test performance more so than retrieval processes associated with recognition (Cooper & Monk, 1976; Darley & Murdock, 1971; Mandler & Rabinowitz, 1981; McDaniel & Masson, 1985; Wenger et al., 1980; see Glover, 1989, for an identical pattern using short texts as the target material). However, there is at present no published work that contrasts testing effects with the types of quizzes commonly found in a classroom (short answer vs. multiple choice) in the context of an actual course with normal classroom content. To address this unexplored issue, in the present study we manipulated the type of quiz in a college Brain and Behavior course at the University of New Mexico. For target facts, we included quizzes with feedback (see Kang, McDermott, & Roediger, 2007 this issue; McDaniel & Fisher, 1991; Pashler, Cepeda, Wixted, & Rohrer, 2005; and Wininger, 2005, for mnemonic advantages of providing feedback after testing) that either required recall (short answer tests) or recognition (multiple choice tests). Generalising from the findings in the experimental memory literature, we predicted that short answer quizzes would produce greater gains in performance on unit exams than would multiple choice quizzes.[1]

An alternative outcome might be possible as well. For the final criterial tasks (unit exams and a cumulative final) we used multiple choice tests, reflecting the kind of assessment test used in most large college classes. With final multiple choice tests, dynamics of transfer appropriate processing may trump the mnemonic benefits of recall over recognition. Transfer appro-

---

[1] Our prediction is based on the observation that the multiple choice questions used herein were worded very similarly or in many cases identically (as for the example quiz item in the Appendix) to the factual statements presented in the textbook. Thus, though multiple choice testing on course content is not necessarily identical to laboratory recognition tests, for the present materials we assume that recognition of the target fact would underlie, at least somewhat, performance on the multiple choice tests.

priate processing refers to increased memory performances when prior processing matches processing required for a subsequent test (see Thomas & McDaniel, in press, for an educationally relevant example, and McDaniel, Friedman, & Bourne, 1978; Morris, Bransford, & Franks, 1977; Roediger & Blaxton, 1987, for basic memory results). On this principle, multiple choice quizzes would presumably promote better transfer to the final multiple choice tests than would the short answer quizzes. (This prediction assumes that recognition processes transfer to subsequent recognition more so than recall processes transfer to subsequent recognition.)

Finally, the testing effect previously reported for educationally relevant materials may represent a somewhat brittle effect that is limited to final criterial questions that are identical to those presented in the initial tests (e.g., Spitzer, 1939, used the same question stems across repeated tests, as did Glover, 1989). The testing effect would be optimally valuable in the classroom if it produced learning of a complex fact, rather than learning of a particular answer when given a particular question. Further, in classroom applications, some instructors are understandably reluctant to give identical questions on quizzes and final assessments. Therefore, in the current study we examined the testing effect in a more challenging setting in which the wording of each question was changed between the initial quiz and the subsequent tests (see Materials section).

METHOD

Participants and design

The participants were 35 male and female students enrolled in a web-based Brain and Behavior course at the University of New Mexico who participated voluntarily for extra credit. All participants completed weekly quizzes, two unit tests, and a final exam that were constructed for the experiment; these tests were not used for evaluation in the course. One participant dropped out of the study before Unit 2; his/her data were not included in any of the analyses. Two additional participants dropped out before the final exam; their data were included in the analyses of quiz and unit test performance, but not for the final exam.

The experiment was a 3×2 within-subjects design, with the quiz type (multiple choice, short answer, read only) and target fact exposure (exposed with quiz/reading, not exposed). As detailed below, a set of not-exposed facts was paired with each exposure condition, yielding a complete factorial design. For purposes of exposition, we have labelled the initial variable "quiz type", though note that one level of the variable (read only) is not quizzed but involves only reading the facts.

Materials

Quizzes.   Two sets of ten facts were extracted from the assigned material that students were instructed to read each week. Weekly reading assignments consisted of approximately 40 pages from an undergraduate textbook (Rosenzweig, Breedlove, & Watson, 2004). Facts selected for this experiment were included in the textbook but were not those that were emphasised in the course itself. For each of the 10 facts in one set, a second fact was taken from the same paragraph to create the second set of facts. For example, the two facts below were taken from the same paragraph of the material and would be assigned to different sets for counterbalancing purposes:

Set 1: All preganglionic axons, whether sympathetic or parasympathetic, release acetylcholine as a neurotransmitter.
Set 2: Parasympathetic postganglionic axons release acetylcholine as a neurotransmitter

During each of 6 weeks one set of facts was quizzed/read and one set was not quizzed/not read. These sets were counterbalanced across participants. In addition, each week across participants each set of facts was presented in one of three "quiz" forms: multiple choice (MC), short answer (SA), or read only (RO). Combining these two counterbalancing factors yielded six counterbalancing groups to which the participants were randomly assigned. Table 1 provides a description of this counterbalancing design and the number of students assigned to each group (see the Appendix for an example of each of the three quiz forms). The experiment spanned 6 weeks of the course, thereby allowing two replications of the design per student.

TABLE 1
Quiz counterbalancing design

| Quiz type | Counterbalancing condition | Fact set | | N |
		A	B	
RO	1	Read	Not read	6
	2	Not read	Read	4
MC	1	Quizzed	Not quizzed	5
	2	Not quizzed	Quizzed	7
SA	1	Quizzed	Not quizzed	7
	2	Not quizzed	Quizzed	5

Counterbalancing conditions have unequal N because some students dropped the course after the initial random assignments to the counterbalancing conditions. RO = read only; MC = multiple choice; SA = short answer.

Feedback. Feedback was constructed for each of the weekly quizzes. Feedback included the test questions, question number (or the read fact), the correct answer, and the participant's response. For the multiple choice questions, all answer choices were given as well. For short answer questions, the question was displayed along with the participant's response and all correct answers. Finally, for read only facts the feedback would always display the fact that was presented on the quiz and a participant response of "I have read the above statement". Examples of all three types of feedback are provided in the Appendix.

Unit tests. Two 60-item multiple choice unit tests were constructed, one for the first 3 weeks of facts presented in the assigned readings and another for the second 3 weeks of facts. The 60 items comprised the entire set of facts quizzed during the previous 3-week period. Note, however, for any particular participant, 30 items had been quizzed (10 MC, 10 SA, 10 RO) and 30 items had not been quizzed (10 yoked to each of the MC, SA, and RO conditions). Feedback was not provided for the unit exams.

To test for retention of the complete conceptual relation (rather than retention of a particular answer provided in a quiz), each fact was tested such that the answer required for a quiz item was now embedded in the question stem, and an alternative portion of the fact was required for the answer. As an example, the quiz wording and unit-test wording for a fact would read as follows:

> Quiz wording: All preganglionic axons, whether sympathetic or para-sympathetic, release _____ as a neurotransmitter:
>   a. acetylcholine
>   b. epinephrine
>   c. norepinephrine
>   d. adenosine

> Unit-test wording: All _____ axons, whether sympathetic or parasympathetic, release acetylcholine as a neurotransmitter.
>   a. preganglionic
>   b. ionotropic
>   c. hypothalamic
>   d. adenosine

Final exam. A multiple choice final exam was constructed. The final exam consisted of all 60 items from both unit tests for a total of 120 items. Half of the items were presented in the same wording as the quiz and half were presented in the same wording as the unit test. Therefore for the final exam, students had seen the exact wording of the question previously.

Procedure

Each week participants were assigned approximately 40 pages of textbook reading in the course. As participants in the research, they were instructed to log on at the end of each week and complete the 10-item quiz over that week's readings. Each week, any particular participant received his or her quiz in a different test format (MC, SA, or RO). On the week when the participant received the RO condition, they simply read the designated target facts and clicked a button for the response "I have read the above statement." Participants were allowed 10 min to complete each quiz; immediately after finishing they were provided access to feedback. The participant clicked on "submit quiz" and received a confirmation statement that their quiz was successfully submitted along with a "view results" link that took them to the results display. The participants could inspect the feedback for as long as they wanted and as many times as they wanted within a week of completing the quiz.

After 3 weeks of quizzes (one MC, one SA, and one RO) the participants were instructed to take the first unit test, with all participants receiving the same unit test. Next, participants were given another 3 weeks of quizzes. Similar to the first 3 weeks, participants were given their quiz in a different test format each week and were provided feedback after each quiz. After completing the second set of three quizzes, participants were instructed to take the second unit test, which tested only the material presented in the second 3 weeks of quizzes. Several weeks after completing the second unit test, participants were instructed to take the final cumulative exam. Students were told that this was a practice cumulative exam that might help them on the in-class final. To avoid contamination, none of the facts tested in the experiment was tested on the actual course exams.[2]

RESULTS AND DISCUSSION

Quiz performance

The mean proportions of quiz questions answered correctly for units one and two are shown in Table 2. A 2×2 within-subjects analysis of variance (ANOVA), with the factors of quiz type (MC or SA) and unit (1 or 2), indicated that there was a main effect of question type such that participants were more likely to answer MC questions correctly than SA questions,

---

[2] Actual course examinations could not be used in the experiment because the Institutional Review Board would not allow the experiment to be conducted as a required part of the course. Consequently, using the material tested in the course as target material for the experiment was judged as possibly coercive and therefore inappropriate.

TABLE 2
Proportion correct on quizzes

	Unit 1	Unit 2
MC	.49 (.21)	.37 (.22)
SA	.17 (.20)	.21 (.20)

Standard deviations are in parentheses. MC = multiple choice; SA = short answer.

$F(1, 33) = 54.12$, $MSE = 0.04$ (for all analyses the alpha level for determining significance was set at .05). There was also a significant interaction between type and unit such that the benefit of answering MC questions compared to SA questions was greater in Unit 1 than in Unit 2, $F(1, 33) = 4.50$, $MSE = 0.04$. The advantage of multiple choice performance over short answer is entirely consistent with the principle that recognition is a less demanding retrieval task than recall. We next examine the extent to which these initial tests influenced performance on subsequent criterial tests.

Unit test performance

The mean proportions of questions answered correctly for the unit one and two multiple choice tests are shown in Table 3. A $2 \times 3 \times 2$ within-subjects ANOVA, with the factors of unit (1 or 2), test type (MC, SA, or RO), and preexposure (quizzed or not quizzed) was conducted on these data. Importantly, there was a main effect of preexposure such that performance was generally better when facts were previously quizzed than when the facts were not quizzed, $F(1, 33) = 14.77$, $MSE = 0.03$. To inform the issues outlined in the introduction, planned contrasts between quizzed and not quizzed items for each quiz type were computed (collapsed across unit; see Figure 1). The advantage of quizzed over not quizzed facts was significant for multiple choice quizzes, $F(1, 64) = 4.00$, $MSE = 0.03$, and for the short answer quizzes, $F(1, 64) = 16.00$, $MSE = 0.03$. There was no

TABLE 3
Proportion of questions answered correctly on Unit 1 and Unit 2 tests

	Unit 1 (N =34)		Unit 2 (N =34)	
	Quizzed	Not quizzed	Quizzed	Not quizzed
MC	.55 (.20)	.50 (.22)	.44 (.25)	.37 (.21)
SA	.61 (.24)	.49 (.21)	.53 (.24)	.42 (.22)
RO	.51 (.22)	.51 (.21)	.43 (.24)	.38 (.19)

Standard deviations are in parentheses. N refers to number of participants. MC = multiple choice; SA = short answer; RO = read only.

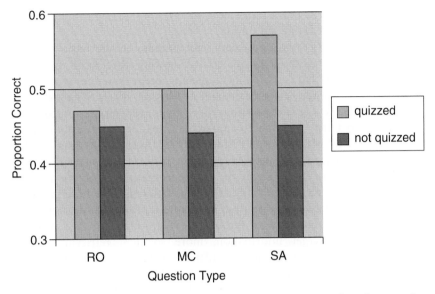

Figure 1.   Unit exam performance of quizzed versus not quizzed items collapsed across units.

significant advantage of presenting facts for reading relative to not presenting the facts ($F < 1$). These patterns clearly reveal a learning benefit of prior quizzing with feedback—a testing effect. Further, this testing effect cannot be interpreted as a mere exposure effect because exposure per se of the facts (the read only condition) produced no significant benefit on the unit test.

There was also a main effect of quiz type such that facts assigned to the SA quiz conditions (exposed and nonexposed) were more accurately learned and retained than facts assigned to either the MC or RO questions, $F(2, 66) = 3.42$, $MSE = 0.03$. Though, the interaction between the factors of test type and preexposure was only marginally significant, $F(2, 66) = 2.55$, $MSE = 0.03$, $p = .07$, examination of Table 3 suggests that the main effect of quiz type was carried by the quizzed (exposed) facts. Planned comparisons confirmed this impression. For exposed facts, there was a significant advantage of short answer quizzing over multiple choice quizzing, $F(1, 66) = 5.44$, $MSE = 0.03$, and read only questions, $F(1, 66) = 11.11$, $MSE = 0.03$, but no significant advantage of multiple choice quizzing relative to reading ($F < 1.0$). For facts with no prior exposure, as expected there were no differences as a function of the particular condition to which the fact was assigned (largest $F < 1.0$). Finally, there was a main effect of unit such that participants performed better on the Unit 1 test than they performed on the Unit 2 test, $F(1, 33) = 18.47$, $MSE = 0.06$. The instructor's impression (JLA) was that students may have been spending less time on the readings midway through the semester (Unit 2 testing) than during the initial part of the semester (Unit 1 testing).

The above results seem consistent with findings in the basic memory literature, using very different materials, showing that recall promotes retrieval processing that is more mnemonically potent than does recognition (Bartlett, 1977; Glover, 1989; McDaniel & Masson, 1985). That is, recall typically produces a greater testing effect than recognition. Indeed, the testing-effect patterns with these complex facts exhibit some similarities to those reported by McDaniel and Masson (1985) using word lists. In that study, cued recall produced significantly better performance on a subsequent cued recall test than did recognition, but importantly half of the time the cues that prompted recall on the final test were different than those that were provided for earlier study and testing. This pattern prompted McDaniel and Masson to suggest that retrieval through recall produces enriched, variable encoding of the target information, more so than retrieval through recognition. The present findings parallel this idea, as cued recall quizzes enhanced performance significantly more than did recognition quizzes on a subsequent test in which the retrieval cues had been altered (i.e., a different question stem was provided than during quizzing).

Interpretation of the potential mechanisms underlying the present effect is more complicated, however. As would be typical for many classroom situations, the present study also provided feedback to students on their quiz responses. Thus, several possible explanations for the over all benefit of SA quizzes over receiving MC quizzes are likely: (a) retrieval failure when recall was attempted (SA) elicited more attentive or effective processing of the feedback than did recognition failure (MC), (b) the SA quiz benefits reflected the potency of recall retrieval relative to recognition retrieval, or (c) both. To gain some insights into these possibilities we conducted a set of conditionalised analyses to attempt to isolate the retrieval and feedback effects of the different quiz types.

Feedback effect. We examined whether the benefits of feedback for missed items were differential across MC and SA quizzes. In addition, we were interested in the general question of whether exposure to the facts as feedback (after missing the fact on a quiz) promoted more learning than exposure to the facts through reading. Accordingly, we calculated the proportion correct on each unit test for items missed on the initial quizzes and compared those values to performance for RO items. A 3 (MC, SA, or RO) × 2 (Unit 1 or Unit 2) within-subjects ANOVA revealed a main effect of quiz type. Examination of Table 4 shows that facts missed on the SA quiz were more likely to be answered correctly on the unit than facts missed on the MC quiz or simply read, $F(2, 64) = 3.18$, $MSE = 0.04$. Planned contrasts

TABLE 4
Proportion of questions answered correctly on unit after being answered incorrectly
on quiz or having been read

	Unit 1 (N =33)	Unit 2 (N =33)
MC	.54 (.23)	.40 (.26)
SA	.61 (.25)	.47 (.24)
RO	.50 (.22)	.43 (.25)

Standard deviations are in parentheses. N refers to number of participants. MC =multiple choice; SA =short answer; RO =read only.

confirmed that missed SA facts were answered correctly (on the unit tests) significantly more often than missed MC facts, $F(1, 64) = 4.90$, $MSE = 0.04$, and significantly more often RO facts, $F(1, 64) = 6.40$, $MSE = 0.04$. The ANOVA also found a main effect of unit such that overall performance was higher for Unit 1 than for Unit 2, $F(1, 32) = 10.77$, $MSE = 0.06$, but there was no interaction of unit with quiz type ($F < 1$).

Note that possible item-selection effects may have contributed to the differential emergence of a positive effect of feedback across missed SA and MC items. Participants missed fewer MC items than SA items, so the missed MC facts were likely relatively harder facts than missed SA facts and certainly harder than the set of entire RO facts. This observation raises two issues. The first is whether positive feedback effects would appear even for missed SA items that were relatively difficult. The second is whether positive feedback effects would appear for missed MC items when a relatively comparable set of RO facts is used as a baseline. In order to examine these issues, we identified a subset of difficult SA questions (questions that were missed on average more than 70% of the time on the initial SA quiz). The resulting subset of difficult SA items consisted of approximately 75% of the original quizzed items. Next we calculated unit test performance sampling only these "difficult" facts, and for each subject only the facts within this set that were answered incorrectly on the quizzes (or all of the difficult facts in the RO condition). The pattern for the "difficult" questions was identical to that found in the initial analysis (thus we dispense with reporting means).

The results are compelling for feedback effects after missing a short answer quiz item. Clearly, learning and retention were better when students were given feedback after missing a short answer question than reading the fact (twice) without being quizzed. In the analyses conducted, the items analysed would overlap considerably across missed SA questions and RO conditions, so that possible item-selection artifacts for this comparison are unlikely. (Item differences would favour the RO condition anyway.) Importantly, the advantage of missed SA items obtained even though the corrected answer on the quiz was not the response that would be required for correct performance on the unit test. Thus, it appears that giving feedback to items not recalled promoted integrated learning of the elements comprising the tested items.

The findings suggest that feedback for missed multiple choice facts did not benefit learning more so than additional exposure (RO). However, because the analysis with the set of difficult facts still could not perfectly equate the set of items compared across MC (only missed items) and RO (all difficult items) presentations, the possibility remains that feedback after missed multiple choice items could produce more learning than RO exposure.

Retrieval effect. To investigate the contribution of retrieval processing to the testing effects for SA and MC questions, we calculated the probability of a correct response on the unit test conditionalised on correct quiz performance. Doing so yielded a robust advantage for facts recovered via recall (SA) than for recognition (MC), but this result is not overly telling (there were far fewer recalled items than recognised items, likely producing item selection effects). Accordingly, rather than report those raw conditionalised results, we instead report the results of a conditionalised analysis based on a limited set of easier items in order to better equate the item difficulty across SA, MC, and RO conditions. The set of easier items were those answered correctly more than 50% of the time on the multiple choice quiz. The analysis is reported for each unit, to minimise deletion of cases due to missing data (e.g., in the SA condition).

The means are shown in Table 5, along with the number of participants who had a complete set of scores (in general, participants with missing data were those who failed to answer any SA questions on the quiz). A within-subjects ANOVA for Unit 1 (with the factors of test type and exposure) found no significant effect. The ANOVA for Unit 2 found a significant effect of quiz type, $F(2, 34) = 5.08$, $MSE = 0.11$. Contrasts confirmed that recalling an answer (SA retrieval) conferred a robust increase in performance rela-

TABLE 5
Proportion of easiest questions answered correctly on the unit test after being answered correctly on the quiz or having been read

	Unit 1 (N = 13)	Unit 2 (N = 18)
MC	.71 (.33)	.58 (.34)
SA	.62 (.46)	.76 (.39)
RO	.55 (.27)	.41 (.34)

Standard deviations are in parentheses. N refers to number of participants. MC = multiple choice; SA = short answer; RO = read only.

tive to reading a fact, $F(1, 34) = 10.21$; recognising an answer (MC retrieval) produced only a marginal advantage relative to reading, $F(1, 34) = 2.41$, $p = .13$.

The Unit 2 results support the idea that retrieval of target information benefits retention more than additional study (the RO condition), with recall rather than recognition-like processes producing the retrieval benefit (see McDaniel & Masson, 1985, for similar findings with word list materials). Further, this pattern also suggests that the overall mnemonic benefit of receiving SA quizzes relative to MC quizzes or to additional presentation of target content (RO condition) was in part due to retrieval effects (at least for the section of the course for which students did not fare as well in general). That is, correct retrieval (recall) on short answer questions appeared to potentiate later test performance.

Final exam performance

The mean proportions for final exam performance were submitted to a $2 \times 2 \times 3 \times 2$ ANOVA, with the factors of unit (1 or 2), quiz preexposure (quiz or no quiz), type of quiz (MC, SA, or RO), and wording (same as quiz or same as unit). The patterns of quiz preexposure evident on the unit test mostly persisted to the final cumulative exam. Performance was generally better for facts exposed in the quiz condition than for facts that were not exposed, $F(1, 31) = 6.14$, $MSE = 0.03$. Critically, there was an interaction between quiz preexposure and quiz type such that the benefit of preexposure significantly varied as a function of quiz type, $F(2, 62) = 4.73$, $MSE = 0.04$. The means representing the interaction are shown in Figure 2.

Planned contrasts of quiz preexposure versus no quiz for each quiz type were calculated to more specifically identify the locus of the interaction. The advantage of quizzed over not quizzed facts was significant for SA quizzes, $F(1, 62) = 6.48$, $MSE = 0.04$, and marginally significant for MC quizzes, $F(1, 62) = 2.88$, $MSE = 0.04$, $p = .09$. RO preexposure produced no benefit relative to no preexposure ($F < 1$). The advantage of quizzed SA items over quizzed MC items that emerged on the unit tests did not reach significance for the final examination performance, $F(1, 62) = 2.00$, $MSE = 0.04$. Additionally, the ANOVA showed a main effect of unit such that overall performance was better for Unit 1 items ($M = 0.54$) than Unit 2 items ($M = 0.47$), $F(1, 31) = 18.87$, $MSE = 0.04$. Finally, there was a significant interaction between unit and quiz preexposure such that a benefit of preexposure was obtained for facts from Unit 2 (M quizzed $= 0.50$, nonquizzed $= 0.44$) but not for facts from Unit 1 (M quizzed $= 0.54$, nonquizzed $= 0.54$), $F(1, 31) = 5.08$, $MSE = 0.03$.

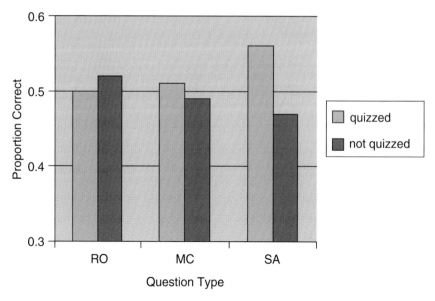

Figure 2.   Final exam performance for quizzed versus not quizzed items.

Test spacing

Because the course was web-based and testing was self-initiated, the dates for which participants logged on to the website were monitored. We calculated how many days apart each participant took the quizzes, unit tests, and final. The means were calculated for the number of days between each consecutive pair of tests and are shown in Table 6. As can be seen from the table, students generally adhered to taking the quizzes about 1 week apart, and took the unit test several days after taking the last quiz. On average, the cumulative final exam was taken about 40 days (5 weeks) after the Unit 2 test, though the range varied from 30 to 56 days.

TABLE 6
Average number of days between tests

Q1–Q2	Q2–Q3	Q3–U1	Q4–Q5	Q5–Q6	Q6–U2	U2–F
7.71	7.76	8.87	4.91	11.00	7.58	40.23

Q = weekly quiz, U = unit test, F = final exam.

CONCLUSIONS

Quizzing improved performance on two unit exams and a cumulative final exam for content covered in a college course relative to content that was not quizzed. Consistent with basic research on the testing effect, the benefit for short answer quizzing was more robust than the benefit for multiple choice quizzing. These findings demonstrate that even in the face of the variable conditions found in a course setting, testing enhances learning and retention (e.g., variability in studying and completing assignments across students, variability in motivation, variability in delays between study and quizzing and between quizzing and criterial testing).

Moreover, the present testing effects were obtained despite the question frames changing from the quizzes to the unit tests. This represents a much more demanding transfer task than implemented in some previous testing effect studies using class-related materials (i.e., text or lectures). In previous studies, the question frames have been repeated across initial and final tests (e.g., Butler & Roediger, 2007 this issue; Glover, 1989; Spitzer, 1939). Thus, the present results demonstrate that learning effects from testing extend beyond mere reproduction of previous quiz answers.

Quizzing that required recall of target information (short answer quizzes), but not quizzing that required recognition (multiple choice quizzes), was more effective than presenting the target information for reading. The present conditions were possibly not optimal for producing the most potent testing effects. In the current experiment, facts were quizzed just once; repeated quizzing for target content produces more robust gains on final tests, even compared against repeated study opportunities (e.g., Roediger & Karpicke, 2006b; Wheeler & Roediger, 1992). Also, as noted above the question frames were not the same across quizzes and unit exams. With repeated quizzing or with similar question frames for quizzes and exam, multiple choice quizzing might become more mnemonically potent than additional exposure.

The present quizzing effects likely depended on feedback being provided for quizzed items (Kang et al., 2007 this issue, provide direct evidence on this point in a laboratory experiment). Conditional analyses of performance on missed items showed that the testing effects were obtained for missed items, at least for the SA quizzes. In light of Pashler et al.'s (2005) report that subsequent performance on missed items not given feedback is very poor, it seems likely that the feedback was instrumental in boosting performance for items not correctly answered on the quiz (again, mainly for short answer questions) (see also Wininger, 2005, for positive benefits of providing feedback on quizzes). This finding raises the interesting question for basic research of why processing feedback of a missed item is more effective than exposure to the target content in the absence of a quiz.

There was also evidence that successful retrieval—primarily that required by SA rather than MC questions—contributed to the present testing effect.[3] Though the analysis of conditionalised performances on which this conclusion is based were not entirely consistent (significant retrieval effects were not found for Unit 1), given the present patterns and previous laboratory findings (e.g., Glover, 1989; McDaniel & Masson, 1985), it seems reasonable that both successful retrieval (recall) and processing of feedback after not being able to retrieve the correct answer contributed to the effective use of short answer quizzing in this experiment.

In closing, the fundamental implication of our findings is that testing to enhance learning should be seriously considered in pedagogical theory and practice. There are compelling strengths of an intervention that uses testing to enhance learning. First, test enhanced learning can be implemented for a variety of course contents. Courses that are heavily fact based, in which students are responsible for learning a large body of facts, seem to be especially good candidates for test enhanced learning. Second, application of test enhanced learning to courses at all levels of the curriculum from primary school to college is straightforward. Though little research has been conducted on testing effects with children, the available work indicates large testing effects for children in elementary school (Metcalfe, Kornell, & Son, 2007 this issue, Exps 1 and 2; Spitzer, 1939). Third, implementing test enhanced learning requires no change in curriculum or teaching style. Indeed, for courses in which web-based assistance is possible, using testing as a learning tool would not require valuable class time. Educational theory and practice would do well not to forget the use of testing as a tool to promote learning and retention.

REFERENCES

Allen, G. A., Mahler, W. A., & Estes, W. K. (1969). Effects of recall tests on long-term retention of paired associates. *Journal of Verbal Learning and Verbal Behavior, 8*, 463–471.

Baine, D. (1986). *Memory and instruction.* Englewood Cliffs, NJ: Educational Technology.

Bartlett, J. C. (1977). Effects of immediate testing on delayed retrieval: Search and recovery operations with four types of cue. *Journal of Experimental Psychology: Human Learning and Memory, 3*, 719–732.

Butler, A. C., & Roediger, H. L., III. (2007). Testing improves long-term retention in a simulated classroom setting. *European Journal of Cognitive Psychology, 19*, 514–527.

Carrier, M., & Pashler, H. (1992). The influence of retrieval on retention. *Memory and Cognition, 20*, 633–642.

---

[3] The mnemonic benefits of retrieving an answer for a short answer quiz question may appear related to the mnemonic benefits of generating a target item during study (i.e., the generation effect, Jacoby, 1978; Slamecka & Graf, 1978). In considering this issue, Carrier and Pashler (1992) suggested, however, that the two effects emerge for different reasons.

Cooper, A. J. R., & Monk, A. (1976). Learning for recall and learning for recognition. In J. Brown (Ed.), *Recall and recognition* (pp. 131–156). New York: Wiley.

Darley, C. F., & Murdock, B. B. (1971). Effects of prior free recall testing on final recall and recognition. *Journal of Experimental Psychology, 91*, 66–73.

Glover, J. A. (1989). The "testing" phenomenon: Not gone but nearly forgotten. *Journal of Educational Psychology, 81*, 392–399.

Hanawalt, N. G., & Tarr, A. G. (1961). The effect of recall upon recognition. *Journal of Experimental Psychology, 62*, 361–367.

Hogan, R. M., & Kintsch, W. (1971). Differential effects of study and test trials on long-term recognition and recall. *Journal of Verbal Learning and Verbal Behavior, 10*, 562–567.

Jacoby, L. L. (1978). On interpreting the effects of repetition: Solving a problem versus remembering a solution. *Journal of Verbal Learning and Verbal Behavior, 17*, 649–667.

Kang, S. H. K., McDermott, K. B., & Roediger, H. L., III. (2007). Test format and corrective feedback modulate the effect of testing on long-term retention. *European Journal of Cognitive Psychology, 19*, 528–558.

Lockhart, R. S. (1975). The facilitation of recognition by recall. *Journal of Verbal Learning and Verbal Behavior, 14*, 253–258.

Mandler, G., & Rabinowitz, J. C. (1981). Appearance and reality: Does a recognition test really improve subsequent recall and recognition? *Journal of Experimental Psychology: Human Learning and Memory, 7*, 79–90.

Masson, M. E. J., & McDaniel, M. A. (1981). The role of organization processes in long-term retention. *Journal of Experimental Psychology: Human Learning and Memory, 7*, 100–110.

Mayer, R. E. (2003). Memory and information processes. In W. M. Reynolds & G. E. Miller (Eds.), *Educational psychology: Vol. 7. Handbook of psychology* (pp. 47–57). Hoboken, NJ: Wiley.

McDaniel, M. A., & Fisher, R. P. (1991). Test and test feedback as learning sources. *Contemporary Educational Psychology, 16*, 192–201.

McDaniel, M. A., Friedman, A., & Bourne, L. E. (1978). Remembering the levels of information in words. *Memory and Cognition, 6*, 156–164.

McDaniel, M. A., Kowitz, M. D., & Dunay, P. K. (1989). Altering memory through recall: The effects of cue-guided retrieval processing. *Memory and Cognition, 17*, 423–434.

McDaniel, M. A., & Masson, M. E. J. (1985). Altering memory representations through retrieval. *Journal of Experimental Psychology: Learning, Memory, and Cognition, 11*, 371–385.

Metcalfe, J., Kornell, N., & Son, L. K. (2007). A cognitive-science based program to enhance study efficacy in a high and low-risk setting. *European Journal of Cognitive Psychology, 19*, 743–768.

Morris, C. D., Bransford, J. D., & Franks, J. J. (1977). Levels of processing versus transfer appropriate processing. *Journal of Verbal Learning and Verbal Behavior, 16*, 519–533.

Pashler, H., Cepeda, N. J., Wixted, J. T., & Rohrer, D. (2005). When does feedback facilitate learning of words? *Journal of Experimental Psychology: Learning, Memory, and Cognition, 31*, 3–8.

Roediger, H. L., III, & Blaxton, T. A. (1987). Retrieval modes produce dissociations in memory for surface information. In D. S. Gorfein & R. R. Hoffman (Eds.), *Memory and learning: The Ebbinghaus Centennial Conference* (pp. 349–379). Hillsdale, NJ: Erlbaum.

Roediger, H. L.III., & Karpicke, J. D. (2006a). The power of testing memory: Basic research and implications for educational practice. *Perspectives on Psychological Science, 1*, 181–210.

Roediger, H. L., III, & Karpicke, J. D. (2006b). Test enhanced learning: Taking tests improves long-term retention. *Psychological Science, 17*, 249–255.

Rosenzweig, M. R., Breedlove, S. M., & Watson, N. V. (2004). *Biological psychology: An introduction to behavioral and cognitive neuroscience*. Sunderland, MA: Sinauer.

Runquist, W. N. (1983). Some effects of remembering on forgetting. *Memory and Cognition, 11*, 641–650.

Slamecka, N. J., & Graf, P. (1978). The generation effect: Delineation of a phenomenon. *Journal of Experimental Psychology: Human Learning and Memory, 4,* 592–604.

Spitzer, H. F. (1939). Studies in retention. *Journal of Educational Psychology, 30,* 641–656.

Thomas, A. K., & McDaniel, M. A. (in press). The negative cascade of incongruent task-test processing. *Memory & Cognition.*

Wenger, S. K., Thompson, C. P., & Bartling, C. A. (1980). Recall facilitates subsequent recognition. *Journal of Experimental Psychology: Human Learning and Memory, 6,* 545–559.

Wheeler, M. A., Ewers, M., & Buonanno, J. F. (2003). Different rates of forgetting following study versus test trials. *Memory, 11,* 571–580.

Wheeler, M. A., & Roediger, H. L., III. (1992). Disparate effects of repeated testing: Reconciling Ballard's (1913) and Bartlett's (1932) results. *Psychological Science, 3,* 240–245.

Whitten, W. B., & Bjork, R. A. (1977). *Learning from tests: Effects of spacing: Journal of Verbal Learning and Verbal Behavior, 16,* 465–478.

Wininger, S. R. (2005). Using your tests to teach: Formative summative assessment. *Teaching of Psychology, 32,* 164–166.

APPENDIX: SAMPLE QUESTIONS AND FEEDBACK

Questions

(MC) All preganglionic axons, whether sympathetic or parasympathetic, release _____ as a neurotransmitter:
    a. acetylcholine
    b. epinephrine
    c. norepinephrine
    d. adenosine

(SA) All preganglionic axons, whether sympathetic or parasympathetic, release _____ as a neurotransmitter.

(RO) All preganglionic axons, whether sympathetic or parasympathetic, release acetylcholine as a neurotransmitter.

Feedback

(MC) All preganglionic axons, whether sympathetic or parasympathetic, release _____ as a neurotransmitter:
    a. acetylcholine
    b. epinephrine
    c. norepinephrine
    d. adenosine
Student Response: b. epinephrine   Correct Answer: a. acetylcholine

(SA) Question 1 All preganglionic axons, whether sympathetic or parasympathetic, release _____ as a neurotransmitter.

Student Response: acetylcholine Correct Answer: acetylcholine

(RO) Question 1 All preganglionic axons, whether sympathetic or parasympathetic, release acetylcholine as a neurotransmitter.

Student Response: I have read the above statement.

EUROPEAN JOURNAL OF COGNITIVE PSYCHOLOGY
2007, 19 (4/5), 514–527

Testing improves long-term retention in a simulated classroom setting

Andrew C. Butler and Henry L. Roediger, III

Washington University in St Louis, St Louis, MO, USA

The benefits of testing on long-term retention of lecture material were examined in a simulated classroom setting. Participants viewed a series of three lectures on consecutive days and engaged in a different type of postlecture activity on each day: studying a lecture summary, taking a multiple choice test, or taking a short answer test. Feedback (correct answers) was provided for half of the responses on the multiple choice and short answer tests. A final comprehensive short answer test was given 1 month later. Restudying or taking a multiple choice test soon after learning improved final recall relative to no activity, but taking an initial short answer test improved final recall the most. Feedback did not affect retention, probably due to the high level of performance on the initial tests. This finding is a powerful demonstration of how tests (especially recall tests) can improve retention of material after long retention intervals.

In most educational settings, tests are employed as a means to evaluate student learning for the purpose of assigning grades. The heavy emphasis on assessment often obscures another function of testing that is highly relevant to the goals of education: the promotion of learning. Considerable research in cognitive psychology has demonstrated that testing improves retention of the material tested, a phenomenon called the testing effect (Carrier & Pashler, 1992; McDaniel & Masson, 1985; Wheeler & Roediger, 1992; see Roediger & Karpicke, 2006a, for a review). To be sure, the idea of using tests as a learning tool in the classroom is not new (Gates, 1917; Jones, 1923–1924; Spitzer, 1939), and many researchers have made a case for the benefit of frequent testing in education (Bangert-Drowns, Kulik, & Kulik, 1991; Foos & Fisher, 1988; Glover, 1989; Leeming, 2002; Paige, 1966). However,

Correspondence should be addressed to Andrew C. Butler, Department of Psychology, Campus Box 1125, Washington University, 1 Brookings Drive, St Louis, MO 63139-4899, USA. E-mail: butler@wustl.edu

We thank Aurora Steinle for her help in creating the experimental materials and collecting data. This research was supported by a grant from the Institute of Education Sciences (No. R305H030339).

DOI: 10.1080/09541440701326097

many of the laboratory studies that demonstrate the benefits of testing utilise basic materials, such as word lists, and retention intervals that usually are quite modest, such as a test at the end of a single experimental session or at most spanning a couple of days (e.g., Allen, Mahler, & Estes, 1969; Hogan & Kintsch, 1971; Thompson, Wenger, & Bartling, 1978). In the effort to apply the benefits of testing to educational practice, an important question remains: to what extent can findings from the laboratory be transferred to the classroom?

Jones (1923–1924) was the first researcher to investigate this question by conducting a series of experiments to study the retention of lecture material in the college classroom. Alarmed by the poor retention of lecture material he found in his first set of experiments (on average only two-thirds of the material was recalled on an immediate test and markedly less after a delay), he decided to assess whether previous findings about the benefits of recitation (Gates, 1917) could be applied to the college classroom. In perhaps his most impressive experiment, Jones investigated the effect of testing on later retention by giving students a brief completion test (e.g., fill-in-the-blank, short answer) immediately after a one hour class lecture and then retesting them after various delays (3 days to 8 weeks) to measure how much of the material they had forgotten. His control condition (for the purpose of comparison with the retest score) was a test of equivalent delay that covered material from the same lecture that had not been previously tested. The data, collected from 600 students across 27 lecture sessions, revealed a large benefit of testing: The amount of information retained after 8 weeks with a prior test was greater than that retained after just 3 days without a prior test. Overall, Jones concluded that testing is an effective method for improving the retention of lecture material and also indicated that tests should be given immediately to maximise their benefit (see also Spitzer, 1939).

The experiments conducted by Jones (1923–1924) are groundbreaking in that he used educationally relevant materials (class lectures) and long retention intervals (up to 8 weeks) to provide solid evidence that tests can be used as learning tools in the classroom. However, one problem with drawing firm conclusions from the results of the study is his failure to equate for total exposure time to the material for the two groups. That is, testing may simply have permitted students to selectively restudy the recalled material, so the benefit from testing could be due just to such restudying. In more recent research on the testing effect, a control group that restudies the material has been employed to equate for overall processing time in order to negate the hypothesis that testing is beneficial only because it involves additional exposure to the material (e.g., Roediger & Karpicke, 2006b). Interestingly, Jones did compare testing to additional study in a separate experiment using paired associates, but chose not to incorporate this design feature in his experiment with class lecture materials (possibly due of the difficulty of producing an appropriate summary of an hour-long lecture).

Since Jones' (1923–1924) landmark study, a handful of subsequent experiments on the testing effect have used complex materials and longer retention intervals, but none has come close to combining the high degree of ecological validity and methodological rigor of his work (but see Metcalfe, Kornell, & Son, 2007 this issue). Many researchers have purposely incorporated educationally relevant materials in carefully controlled experiments with the goal of generalising to the classroom, a practice that dates from some of the first studies demonstrating the testing effect (e.g., Gates, 1917; Spitzer, 1939) to more recent efforts that have revived this tradition (e.g., Roediger & Karpicke, 2006b). Among the types of materials that have been used are foreign language paired associates (Carrier & Pashler, 1992), general knowledge questions (McDaniel & Fisher, 1991; Butler, Karpicke, & Roediger, 2007), and prose passages (Duchastel & Nungester, 1981; Foos & Fisher, 1988; Glover, 1989; LaPorte & Voss, 1975; Roediger & Karpicke, 2006b). Although the use of complex verbal materials in laboratory studies has strengthened the rationale for applying testing as a learning tool in education, even the most complex verbal materials (e.g., prose passages) are still relatively simple compared to the rich array of information encountered by students in the classroom.

Investigations of the testing effect that incorporate a retention interval of more than a week are rare and, to our knowledge, almost all of these studies have utilised naturalistic methodology to examine the extent to which information is retained over long periods of time. A prime example is the literature on the long-term retention of knowledge acquired in classrooms (e.g., Landauer & Ainsle, 1975; Semb, Ellis, & Araujo, 1993; for review see Semb & Ellis, 1994). Taken as a whole, these studies suggest that classroom testing benefits long-term retention of course material across a range of disciplines (e.g., medical education, physics, language instruction, etc.). However, instead of manipulating testing as an independent variable, these studies use testing to examine the retention of information over the period between a final course exam and a subsequent retention exam (often given as an afterthought) as a function of other variables, such as instructional technique and degree of original learning. In addition, these studies were conducted in real classrooms using established curriculum, a situation that introduces numerous uncontrolled factors (e.g., studying outside the classroom) and a lack of random assignment to groups (because ethical objections about placing students in a true control group). Another relevant example is research that investigates the maintenance of knowledge over retention intervals of many years. Bahrick and his colleagues have produced some of the best research on this topic showing the long-lasting benefits of testing (Bahrick, 1979; Bahrick, 1984; Bahrick & Hall, 1991). However, one limitation of his studies is that he must rely on estimations of original learning in order to make feasible decade-long retention intervals. A cross-sectional design has also been used to study the long-term retention of knowledge learned in a cognitive psychology course (Conway, Cohen, & Stanhope, 1991).

Of the few studies that have manipulated testing as an independent variable to examine retention over longer intervals, almost all have used the relatively simple verbal materials described above (e.g., Nungester & Duchastel, 1982; Spitzer, 1939). One notable exception is a recent study by McDaniel, Anderson, Derbish, and Morrisette (2007, this issue) that investigated the benefits of testing over a semester using complex verbal materials. Students in a web-based course on "Brain and Behavior" were assigned 40 pages of reading per week and took either a short answer quiz, a multiple choice quiz, or read the facts that were used for the quiz conditions. Taking an initial short answer quiz led to superior performance on a subsequent multiple choice unit test relative to taking an initial multiple choice quiz or reading key facts.

The present experiment attempts to build upon the earlier work of Jones (1923–1924) by investigating the benefits of testing in a simulated college classroom setting. The study combined the experimental control of the laboratory with materials (art history lectures) like those found in a college classroom. We also used a long retention interval (1 month) to provide insight into a more realistic timescale over which students may retain classroom lecture information prior to a test. In addition to incorporating a "study" control group to equate presentation with the testing groups for total exposure to the materials, we investigated how different types of test (multiple choice and short answer) and the provision of feedback (correct answers given or not) would benefit retention of lecture material.

Participants watched a series of three lectures on consecutive days and engaged in a different learning activity after each lecture: taking a multiple choice test, taking a short answer test, or studying a lecture summary that contained points tested in other conditions. Each learning activity incorporated information from the lecture viewed that day only. Correct answer feedback (a presentation of the question and correct response) was given for half of the responses on the multiple choice and short answer tests.

One month later, participants returned for a comprehensive short answer exam that covered all three lectures. This final test included questions about information covered in the learning activities as well as information that had appeared in the lectures but that had not been re-presented during the any of the learning activities. This material from the lecture that was not presented again in any condition serves as a baseline against which to assess the effects of restudying or taking a multiple choice or short answer test.

METHOD

Participants and design

Twenty-seven Washington University undergraduates participated in the experiment (six other participants completed the initial sessions, but chose not to return for the final session and were therefore replaced by new participants). Course credit was given for the initial three learning sessions and a payment of $10 was given for the final test session. Participants were tested in groups of two to six people. The experiment employed a 2 (type of postlecture activity: multiple choice, short answer) $\times$ 3 (provision of test/feedback: no test, test without feedback, test with feedback) within-participants design. We also included an additional study control condition (a third type of postlecture activity) that could not be crossed with the provision of test/feedback factor. Thus, the overall design was unbalanced, but the experiment was fully counterbalanced and utilised a completely within-participants design. The type of postlecture activity factor and the additional study control were manipulated between lectures, whereas the provision of test/feedback factor was manipulated within lectures, but between items. That is, for each lecture in a testing condition, 10 items were not tested, 10 items were tested without feedback, and 10 items were tested with feedback.

Materials

Materials consisted of three videotaped lectures on art history from a series entitled *From Monet to Van Gogh: A history of impressionism* (The Teaching Company, 2000). The videos depicted a professor (Dr Richard Brettel) lecturing into the video camera (as if speaking to a classroom of students) interspersed with slides of relevant pieces of art and photographs. Each lecture covered the life and work of a single artist (Berthe Morisot, Auguste Renoir, Edgar Degas) and lasted 30 min.

For the purpose of the postlecture activities, 30 facts were selected from each lecture to create study and test materials. These facts covered many types of information (e.g., names, dates, events, etc.) and the timing of their presentation during the course of the lecture was evenly distributed over the 30 min. Lecture summary materials (for the study condition) were constructed by grouping the facts into paragraphs. Test materials were constructed by converting the facts into question/answer format. For example, a question from the Morisot lecture was "What aspect of Morisot's art could be used to date her paintings?" (Answer: *The fashions worn by the women*). For the purpose of multiple choice test, three plausible lures were developed for each question.

The experiment was counterbalanced in several ways. First, the 30 facts/ questions for each lecture were divided into three sets of 10 items: Sets A, B, and C. To create the sets, the facts/questions were arranged by order of presentation in the lecture and randomly assigned to a set with the constraint that no set could receive more than one item from each consecutive group of three items. This method ensured that each set contained items that were evenly distributed over the course of the lecture. Second, three lecture presentation orders were created to counterbalance the sequence in which participants would view the lectures in the three initial learning sessions. The three orders were constructed such that overall each lecture would be presented equally often in each presentation position: (1) Renoir/Morisot/Degas, (2) Degas/Renoir/Morisot, (3) Morisot/Degas/ Renoir. Third, three orders of the postlecture learning activities were created to counterbalance the sequence in which participants would engage in the different tasks. These orders were established such that across participants each activity occurred equally often after each session: (1) multiple choice/ short answer/study, (2) study/multiple choice/short answer, (3) short answer/ study/multiple choice. Finally, the counterbalancing orders for item set, lecture, and postlecture learning activity were factorially combined to create a total of 27 versions of the experiment. Each of the 27 participants was randomly assigned to one of these 27 versions.

Procedure

The experiment consisted of three initial learning sessions, which occurred on successive days, and a final test session, which took place about 1 month (28 days) after the final learning session. None of participants reported any prior experience with the material (e.g., an art history course on Impressionism).

Initial learning sessions.   At the first session, participants were given a general overview of the experiment. Before watching the video, they were instructed to approach the lecture as they would a regular class and to take notes on blank paper that was provided. Although each participant took notes during all three initial sessions, the instruction to take notes was included to enhance the simulation of a classroom experience and therefore the notes were not subjected to any further analysis. When everyone was ready to begin, the video lecture was presented on a large screen at the front of the room by way of a mounted projector. After the lecture, the participants handed in their notes and moved to a computer to engage in the postlecture learning activity: studying a summary of the lecture, taking a multiple choice test, or taking a short answer test (depending on the task to which they were assigned for that session). All the postlecture learning activities were presented individually on a PC using E-Prime software (Schneider, Eschman, & Zuccolotto, 2002) and the specific instructions for the assigned activities were explained at the start of the computer program. The postlecture portion of the session lasted approximately 10 min. Both the multiple choice and short answer tests were self-paced and the 20 questions were presented in a random order determined by the program (the other 10 questions associated with the lecture were not tested). Before the short answer test, participants were instructed to provide an answer to every question and told that any given answer should be no more than a sentence in length. Responses were entered using the keyboard. On both types of test, participants rated the confidence in their response after each question on a 4-point scale: $0 = $ guess, $1 = $ low confidence, $2 = $ medium confidence, or $3 = $ high confidence. After the confidence rating, participants saw either the correct answer feedback or a screen with "loading next question" for 6 s after each question (depending on the condition to which the item was assigned), so that total time spent on each question was roughly equated.

The study task consisted of reading a summary of the lecture that included all 30 facts from the lecture. Participants were instructed to read through the summary and pick up any information they had missed in the lecture. For the purpose of presentation, the summary was split up into three sections. Each section was displayed for 90 s (sufficient time to read through text once) before the program automatically cycled on to the next section. In total, the summary was presented twice (two complete cycles of the three sections) to keep participants engaged for the full duration of the postlecture activity. Thus, the time spent on each of the different postlecture activities was roughly equated with each task lasting approximately 10 min. The subsequent two learning sessions followed the same format: Participants watched a lecture (30 min) and then engaged in one of three postlecture learning activities (study, multiple choice test, short answer test). At the end of the third learning session, they were reminded about the final session and dismissed.

Final test session. Approximately 1 month after the third learning session, participants returned to take the comprehensive, self-paced, short answer test. The test consisted of 90 questions and covered all three lectures. As before, the test was given on a PC computer and the questions were presented in random order. Responses were entered using the keyboard. Instructions against guessing were given ("please answer only if you are reasonably sure you are correct") and thus omitting a response was identified as an option. After participants had finished the final test, they were debriefed and dismissed.

RESULTS

All results were significant at the .05 level of confidence unless otherwise noted. Pairwise comparisons were Bonferroni-corrected to the .05 level. In the analysis of repeated measures, a Geisser-Greenhouse correction was used for violations of the sphericity assumption (Geisser & Greenhouse, 1958).

Initial learning tests: Proportion correct

Overall, participants produced a high level of initial test performance: the proportion of correct responses on the multiple choice test ($M = 0.88$) was significantly higher than that of the short answer test ($M = 0.68$). However, this high level of performance was intended for two reasons: (1) to make sure that performance on the final test would be above floor, and (2) when using a test as a learning tool it is important that test-takers are able to retrieve a reasonable amount of the tested information, as Jones (1923–1924) and others have pointed out previously.

Final short answer test: Proportion correct

Figure 1 shows the proportion of correct recall for the final short answer test as a function of initial learning activity condition (data in the test conditions are collapsed across feedback conditions). The mean proportion correct for items in the no feedback and feedback conditions were almost identical for both types of prior test: multiple choice (no feedback = .36, feedback = .36)

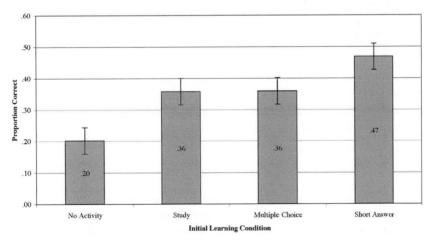

Figure 1. Mean proportion correct recall on the final short answer test as a function of initial postlecture learning condition (errors bars represent 95% confidence intervals).

and short answer (no feedback = .46, feedback = .47). This observation was confirmed by a 2 (initial test type: multiple choice, short answer) × 2 (provision of feedback: no feedback, feedback) repeated measures ANOVA in which there was no difference between provision of feedback conditions, $F(1, 26) = 0.01$, $MSE = 0.019$, $p = .95$. There was a significant main effect of initial test type, $F(1, 26) = 15.96$, $MSE = 0.020$, partial $\eta^2 = .38$, where taking a prior short answer test ($M = 0.47$) led to superior performance relative to a prior multiple choice test ($M = 0.36$). The interaction of initial test type and provision of feedback was not significant, $F(1, 26) = 0.09$, $MSE = 0.023$, $p = .76$. Thus, for the purpose of subsequent analysis, the data in the prior testing conditions were collapsed across the feedback conditions.

To examine the relative benefit of prior learning activity, a one-way repeated measures ANOVA was conducted with type of initial learning task (no test, study, multiple choice, short answer) as the factor and proportion correct as the dependent variable. This test revealed a significant difference among the four initial learning task conditions, $F(3, 78) = 27.07$, $MSE = 0.012$, partial $\eta^2 = .51$. Pairwise comparisons indicated that a higher proportion of items in the short answer condition were recalled than in either the multiple choice condition, $t(26) = 3.99$, $SEM = 0.027$, or the study condition, $t(26) = 3.17$, $SEM = 0.035$. There was no difference between the multiple choice and study conditions, $t(26) = .04$, $SEM = 0.031$, $p = .97$, but the study condition (and the other conditions) led to a higher proportion of correct responses than the no test condition, $t(26) = 5.10$, $SEM = 0.031$.

Final short answer test: Performance as a function of initial confidence

Performance for the two initial test conditions (multiple choice and short answer) was broken down as a function of initial confidence estimates. Due to the high level of initial test performance, confidence estimates were skewed towards the "high confidence" end of the scale. Overall, higher levels of initial confidence led to a higher proportion correct on the final short answer test. However, there were no systematic differences between the two feedback conditions at any of confidence levels.

GENERAL DISCUSSION

This experiment examined how different types of postlecture activity affected retention of lecture material over a realistic (1 month) retention interval as measured by a final short answer test. We found that taking a prior short answer test produced significantly better retention of the material than both studying a lecture summary or taking a multiple choice test.

Although there was no difference in the amount of material retained in the study and multiple choice conditions, all three conditions in which participants engaged in a postlecture activity resulted in superior performance relative to the no activity condition. Surprisingly, the provision of feedback after responses did not improve retention of the material in either of the test activity conditions (multiple choice and short answer). We now turn to discussing each of these results.

The primary finding was that taking a short answer test produced superior retention of lecture material after 1 month relative to studying a lecture summary, a control condition in which participants were essentially shown twice all the critical facts that would later be tested. This result provides compelling evidence that testing can improve the retention of classroom lecture material by way of a postlecture test procedure that can be easily implemented in the classroom. Many studies have found that taking a test leads to greater retention of the material relative to a restudy condition, especially when the test involves response production (e.g., Duchastel & Nungester, 1981; Hogan & Kintch, 1971; McDaniel et al., 2007 this issue; Roediger & Karpicke, 2006b; Thompson et al., 1978).

The short answer test condition also produced superior retention relative to the multiple choice condition. This result fits well with previous laboratory research using basic materials that shows that taking an initial recall test confers larger benefits on subsequent test than taking an initial recognition test (Cooper & Monk, 1976; McDaniel & Masson, 1985; Wenger, Thompson, & Bartling, 1980). This result is also consistent with research using educationally relevant materials in which an initial short answer test produces superior performance on a subsequent test relative to an initial multiple choice test (Duchastel, 1981; Kang, McDermott, & Roediger, 2007 this issue; McDaniel et al., 2007 this issue), a result that generally occurs regardless of whether the final test is in short answer or multiple choice format (e.g., Foos & Fisher, 1988; Glover, 1989; Kang et al., 2007 this issue). Some studies have found an overall superiority of initial multiple choice test, but this may be due to very low performance on the initial short answer test (e.g., Kang et al., 2007 this issue, Exp. 1). Theoretically, one explanation for these results is the idea that greater depth or difficulty in retrieval leads to better retention of the information tested (Bjork, 1975; McDaniel & Masson, 1985). Presumably, short answer tests (which require the production of a response) involve a greater degree of retrieval difficulty than multiple choice tests (which require the selecting the correct response from a number of alternatives). In addition, the results could be explained within a transfer-appropriate processing framework (Morris, Bransford, & Franks, 1977). Taking an initial short answer test would be expected to promote better transfer relative to an initial multiple choice test when the final test is a short answer test. It is important to note that these two theoretical explanations are not mutually exclusive, and both likely play a role in producing the present results.

The finding that the short answer test condition produced better retention than the no activity control condition is notable in that it replicates the findings of Jones (1923–1924) as well as other previous studies (Duchastel, 1980; Glover, 1989; Wheeler & Roediger, 1992). Interestingly, performance on the final short answer test was equivalent for the multiple choice and study groups. A possible explanation is that the lecture summary provided in the study condition gave a distinct advantage by exposing participants twice to all the critical facts that they would later be tested on. With respect to educational practice, this is a rather artificial study task because educators would never give students the answers to the test ahead of time. A more realistic control condition, and one we recommend for future studies of this type, would be to permit students to review the notes they took during the lecture.

One puzzling finding is that feedback did not improve retention of the material. In many experiments, feedback has a profound effect on retention (e.g., McDaniel & Fisher, 1991) because it helps test-takers to correct errors (Bangert-Drowns, Kulik, Kulik, & Morgan, 1991; Pashler, Cepeda, Wixted, & Rohrer, 2005) and to confirm correct responses (Butler et al., 2007). This null effect is likely due in part to the high level of performance on the initial tests, especially for multiple choice: Feedback was not as useful because few errors were made (see Kang et al., 2007 this issue). However, this reasoning cannot fully explain why feedback did not have an effect on the initial short answer test as participants got almost a third of the responses incorrect ($M = 0.32$). Other factors that may have led to ineffectiveness of feedback were the amount of time participants were given to process the feedback and the fact that it occurred immediately after subjects responded. The information tested by any given question was quite complicated (e.g., an answer often consisted of a long phrase or sentence). Feedback was presented for only 6 s and this amount of time may not have been sufficient to allow participants to fully process the information. The timing of feedback may be critical, because evidence exists suggesting that the ratio between the interstudy interval and retention interval maximises retention (Cepeda, Pashler, Vul, Wixted, & Rohrer, 2006). If feedback is conceptualised as an additional study opportunity (i.e., in addition to the initial exposure to the material), this research would suggest that should be presented after a delay in order to produce spaced presentations and optimal retention. Of course, an alternative hypothesis is that giving immediate feedback simply does not affect retention over long periods of time, but this generalisation seems unlikely because the type of feedback used (a re-presentation of the question and the correct answer) almost always increases learning from tests (see Roediger & Karpicke, 2006a). On a related note, feedback may be very important to reducing the negative effects that arise from exposing test-takers to misinformation in the form of multiple choice lures (Roediger & Marsh, 2005). However, in our experiment, very few lure items from the multiple choice test were produced as answers on the final short answer test ($M = 0.04$), and the proportion of lures produced on the final test in the initial multiple choice condition did not differ from the baseline rate of spontaneously producing these responses in the other conditions.

We believe the present findings have direct implications for educational practice. Our experiment combined ecologically valid presentation materials (actual lectures) and realistic retention intervals (1 month). This combination makes our study one of the most powerful demonstrations to date of how the mnemonic benefits of testing can be applied to enhance classroom learning. The benefit of taking a brief quiz (either short answer or multiple choice) is especially striking when compared with the no activity condition, which is perhaps more indicative of common practice in the classroom than the restudy condition. In addition to boosting retention, frequent testing can help to lower students' test anxiety and increase the regularity of studying (Leeming, 2002). Although it did not have an effect in the present study, feedback should also be provided to ensure students are learning from the test, especially in the event of poor test performance. To minimise the time taken away from the primary classroom activities, feedback could be accomplished by requiring students to self-correct their tests after the class period. We encourage educators to incorporate testing into their daily classroom routine: The amount of class time sacrificed for a quiz is small compared to gain in retention of material.

REFERENCES

Allen, G. A., Mahler, W. A., & Estes, W. K. (1969). Effects of recall tests on long-term retention of paired associates. *Journal of Verbal Learning and Verbal Behavior, 8*, 463–470.

Bahrick, H. P. (1979). Maintenance of knowledge: Questions about memory we forgot to ask. *Journal of Experimental Psychology: General, 108*, 296–308.

Bahrick, H. P. (1984). Semantic memory content in permastore: 50 years of memory for Spanish learned in school. *Journal of Experimental Psychology: General, 113*, 1–29.

Bahrick, H. P., & Hall, L. K. (1991). Lifetime maintenance of high school mathematics content. *Journal of Experimental Psychology: General, 120*, 20–33.

Bangert-Drowns, R. L., Kulik, J. A., & Kulik, C. C. (1991). Effects of frequent classroom testing. *Journal of Educational Research, 85*, 89–99.

Bangert-Drowns, R. L., Kulik, C. C., Kulik, J. A., & Morgan, M. (1991). The instructional effect of feedback in test-like events. *Review of Educational Research, 61*, 213–238.

Bjork, R. A. (1975). Retrieval as a memory modifier: An interpretation of negative recency and related phenomena. In R. L. Solso (Ed.), *Information processing and cognition* (pp. 123–144). New York: Wiley.

Butler, A. C., Karpicke, J. D., & Roediger, H. L. III., (2007). A matter of confidence: Correct responses benefit from feedback. *Manuscript submitted for publication.*

Carrier, M., & Pashler, H. (1992). The influence of retrieval on retention. *Memory and Cognition, 20,* 633–642.

Cepeda, N. J., Pashler, H., Vul, E., Wixted, J. T., & Rohrer, D. (2006). Distributed practice in verbal recall tasks: A review and quantitative synthesis. *Psychological Bulletin, 132,* 354–380.

Conway, M. A., Cohen, G., & Stanhope, N. (1991). On the very long-term retention of knowledge acquired through formal education: Twelve years of cognitive psychology. *Journal of Experimental Psychology: General, 120,* 395–409.

Cooper, A. J. R., & Monk, A. (1976). Learning for recall and learning for recognition. In J. Brown (Ed.), *Recall and recognition* (pp. 115–140). London: Wiley.

Duchastel, P. C. (1980). Extension of testing effects on the retention of prose. *Psychological Reports, 47,* 1062.

Duchastel, P. C. (1981). Retention of prose following testing with different types of test. *Contemporary Educational Psychology, 6,* 217–226.

Duchastel, P. C., & Nungester, R. J. (1981). Long-term retention of prose following testing. *Psychological Reports, 49,* 470.

Foos, P. W., & Fisher, R. P. (1988). Using tests as learning opportunities. *Journal of Educational Psychology, 80,* 179–183.

Gates, A. I. (1917). Recitation as a factor in memorizing. *Archives of Psychology, 6,* No. 40.

Geisser, S., & Greenhouse, S. W. (1958). An extension of Box's results on the use of F distribution in multivariate analysis. *Annals of Mathematical Statistics, 29,* 885–891.

Glover, J. A. (1989). The "testing" phenomenon: Not gone but nearly forgotten. *Journal of Educational Psychology, 81,* 392–399.

Hogan, R. M., & Kintsch, W. (1971). Differential effects of study and test trials on long-term recognition and recall. *Journal of Verbal Learning and Verbal Behavior, 10,* 562–567.

Jones, H. E. (1923). The effects of examination on the performance of learning. *Archives of Psychology, 10,* 1–70.

Kang, S. H. K., McDermott, K. B., & Roediger, H. L., III. (2007). Test format and corrective feedback modulate the effect of testing on long-term retention. *European Journal of Cognitive Psychology, 19,* 528–558.

Landauer, T. K., & Ainslie, K. I. (1975). Exams and use as preservatives of course-acquired knowledge. *Journal of Educational Research, 69,* 99–104.

LaPorte, R. E., & Voss, J. F. (1975). Retention of prose materials as a function of postacquisition testing. *Journal of Educational Psychology, 67,* 259–266.

Leeming, F. C. (2002). The exam-a-day procedure improves performance in psychology classes. *Teaching of Psychology, 29,* 210–212.

McDaniel, M. A., Anderson, J. L., Derbish, M. H., & Morrisette, N. (2007). Testing the testing effect in the classroom. *European Journal of Cognitive Psychology, 19,* 494–513.

McDaniel, M. A., & Fisher, R. P. (1991). Tests and test feedback as learning sources. *Contemporary Educational Psychology, 16,* 192–201.

McDaniel, M. A., & Masson, M. E. J. (1985). Altering memory representations through retrieval. *Journal of Experimental Psychology: Learning, Memory and Cognition, 11,* 371–385.

Metcalfe, J., Kornell, N., & Son, L. K. (2007). A cognitive-science based programme to enhance study efficacy in a high- and low-risk setting. *European Journal of Cognitive Psychology, 19,* 743–768.

Morris, P. E., Bransford, J. D., & Franks, J. J. (1977). Levels of processing versus transfer-appropriate processing. *Journal of Verbal Learning and Verbal Behavior, 16,* 519–533.

Nungester, R. J., & Duchastel, P. C. (1982). Testing versus review: Effects on retention. *Journal of Educational Psychology, 74,* 18–22.

Paige, D. D. (1966). Learning while testing. *Journal of Educational Research, 59*(6), 276–277.

Pashler, H., Cepeda, N. J., Wixted, J. T., & Rohrer, D. (2005). When does feedback facilitate learning of words? *Journal of Experimental Psychology: Learning, Memory and Cognition, 31*, 3–8.

Roediger, H. L., III, & Karpicke, J. D. (2006a). The power of testing memory: Basic research and implications for educational practice. *Perspectives on Psychological Science, 1*, 181–210.

Roediger, H. L., III, & Karpicke, J. D. (2006b). Test-enhanced learning: Taking memory tests improves long-term retention. *Psychological Science, 17*, 249–255.

Roediger, H. L., III, & Marsh, E. J. (2005). The positive and negative consequence of multiple-choice testing. *Journal of Experimental Psychology: Learning, Memory and Cognition, 31*, 1155–1159.

Schneider, W., Eschman, A., & Zuccolotto, A. (2002). *E-Prime reference guide.* Pittsburgh, PA: Psychology Software Tools, Inc.

Semb, G. B., & Ellis, J. A. (1994). Knowledge taught in school: What is remembered? *Review of Educational Research, 64*, 253–286.

Semb, G. B., Ellis, J. A., & Araujo, J. (1993). Long-term memory for knowledge learned in school. *Journal of Educational Psychology, 85*, 305–316.

Spitzer, H. J. (1939). Studies in retention. *Journal of Educational Psychology, 30*, 641–656.

The Teaching Company (Producer). (2000). *From Monet to Van Gogh: A history of impressionism* [Motion picture]. (Available from The Teaching Company, Chantilly, VA)

Thompson, C. P., Wenger, S. K., & Bartling, C. A. (1978). How recall facilitates subsequent recall: A reappraisal. *Journal of Experimental Psychology: Human Learning and Memory, 4*, 210–221.

Wenger, S. K., Thompson, C. P., & Bartling, C. A. (1980). Recall facilitates subsequent recognition. *Journal of Experimental Psychology: Human Learning and Memory, 6*, 545–559.

Wheeler, M. A., & Roediger, H. L., III. (1992). Disparate effects of repeated testing: Reconciling Ballard's (1913) and Bartlett's (1932) results. *Psychological Science, 3*, 240–245.

EUROPEAN JOURNAL OF COGNITIVE PSYCHOLOGY
2007, 19 (4/5), 528–558

Test format and corrective feedback modify the effect of testing on long-term retention

Sean H. K. Kang, Kathleen B. McDermott, and
Henry L. Roediger, III

Washington University in St Louis, St Louis, MO, USA

We investigated the effects of format of an initial test and whether or not students received corrective feedback on that test on a final test of retention 3 days later. In Experiment 1, subjects studied four short journal papers. Immediately after reading each paper, they received either a multiple choice (MC) test, a short answer (SA) test, a list of statements to read, or a filler task. The MC test, SA test, and list of statements tapped identical facts from the studied material. No feedback was provided during the initial tests. On a final test 3 days later (consisting of MC and SA questions), having had an intervening MC test led to better performance than an intervening SA test, but the intervening MC condition did not differ significantly from the read statements condition. To better equate exposure to test-relevant information, corrective feedback during the initial tests was introduced in Experiment 2. With feedback provided, having had an intervening SA test led to the best performance on the final test, suggesting that the more demanding the retrieval processes engendered by the intervening test, the greater the benefit to final retention. The practical application of these findings is that regular SA quizzes with feedback may be more effective in enhancing student learning than repeated presentation of target facts or taking an MC quiz.

A wealth of empirical research has found that retention of studied material can be enhanced by testing. A memory test does not merely measure the amount of learning; it has an impact on the state of that memory itself (Lachman & Laughery, 1968; Tulving, 1967). Subjects who

Correspondence should be addressed to Sean Kang, Department of Psychology, Washington University, St Louis, MO 63130-4899, USA. E-mail: seankang@wustl.edu

This research was conducted as part of the Master's Thesis of the first author, and was supported by a grant from the Institution of Education Sciences (R305H030339) to the second and third authors. The results from Experiment 2 were presented as a poster at the 17th annual convention of the American Psychological Society, Los Angeles, CA, in May 2005. The authors acknowledge Mark McDaniel for his valuable comments on an earlier version of this manuscript. Appreciation is also extended to Seth Goodman for assistance with subject testing and data entry.

receive an intervening test after the initial learning experience typically perform better on a later final test, relative to subjects given only the final test. This phenomenon has come to be referred to as the *testing effect*, and has been demonstrated with diverse study stimuli, including word lists (Darley & Murdock, 1971), paired associates (Runquist, 1986), pictures (Wheeler & Roediger, 1992), general knowledge facts (McDaniel & Fisher, 1991), and prose passages (LaPorte & Voss, 1975; for a review, see Roediger & Karpicke, 2006a). One question that remains to be resolved is how the format of the initial test influences the testing effect. Do multiple choice (MC) or short answer (SA) tests differ in the benefit they produce on a final test? Does the answer depend on the format of the final test? This issue is important for both practical and theoretical reasons.

EDUCATIONAL RELEVANCE

In education, tests have typically been employed solely as assessment tools for evaluating learning and academic progress (Dempster, 1996). High-stakes achievement tests have acquired a less than desirable reputation in recent years, with critics charging that such tests are culturally biased and encourage instructors to devote too much time teaching to the test (Anderson, 1998). One concern is that this may result in a spillover effect, such that educators develop an aversion to all types of testing. The fact overlooked in this debate is that tests can be utilised as instruments to promote learning and retention, as demonstrated by Spitzer (1939), who tested the entire sixth-grade population of 91 elementary schools in Iowa. After students read a paper, he varied the number of tests and the retention intervals between tests, and found that students who were tested soon after reading the paper retained the material better on a test given 63 days later. Despite this and other early studies yielding similar findings (Gates, 1917; Jones, 1923), this benefit of testing has remained largely untapped by educators, with the research findings rarely communicated in teacher education courses or implemented in pedagogical practice (Dempster & Perkins, 1993). A major goal of educators is the long-term retention of knowledge acquired by students (Halpern & Hakel, 2002). The current study is an effort to provide evidence-based recommendations for the use of testing to enhance learning and retention, by using more educationally relevant materials than word lists. To discover the type of testing that may be best for implementation in educational settings, our study compared two testing formats that are commonly used in classrooms—MC and SA—to ascertain which type of format would be more beneficial for later retention. An MC test involves recognition: The subject has to discriminate among the options provided in order to choose the answer. An SA test, on the other hand, involves response production: The subject must retrieve and generate an answer in response to the question cue.

THEORETICAL ISSUES

In addition to the practical issues surrounding the testing effect, we were also interested in examining the theoretical mechanisms that underlie it. One theoretical account is that the engagement of retrieval processes during an initial test modifies the memory trace of target items (Bjork, 1975), increasing the probability of successful retrieval later. Many studies have shown that receiving a test, relative to having no test, improves later retention (e.g., Darley & Murdock, 1971; Runquist, 1983, 1986, 1987). However, this sort of comparison leaves open the possibility that it is the re-presentation of an item which occurs during the test, rather than the act of retrieval per se, that enhances retention. Indeed, from the results of a multitrial free recall experiment in which the number and sequence of study and test trials were varied, Tulving (1967) concluded that study and test trials seem to facilitate subsequent recall to the same extent. Other researchers have also argued that a presentation at test (i.e., successfully recalling an item) is functionally similar to a study presentation, and that any effect of testing is due to an overlearning of a subset of items (Slamecka & Katsaiti, 1988; Thompson, Wenger, & Bartling, 1978). More recent research, however, has shown that the testing effect cannot be wholly accounted for by the amount of exposure to the tested material, since the testing effect is still obtained when memory for tested items is compared to memory for items that are re-presented but not tested (Carrier & Pashler, 1992; Kuo & Hirshman, 1996; Roediger & Karpicke, 2006b; Wheeler, Ewers, & Buonanno, 2003).

According to the transfer appropriate processing framework (Blaxton, 1989; Morris, Bransford, & Franks, 1977), memory performance depends on the overlap between encoding and retrieval processes. Applied to the current context, this framework provides an alternative explanation for the testing effect: It is the engagement of similar operations on the intervening and final tests that results in better performance for previously tested items, relative to items that were not initially tested or only restudied.

Several studies have examined the issue of how test format affects later memory performance. It has been demonstrated that taking a test of a particular format can still lead to a positive transfer to a later test of a different format: For example, prior recall tests facilitate subsequent recognition (Hanawalt & Tarr, 1961; Lockhart, 1975; Wenger, Thompson, & Bartling, 1980), prior recognition tests facilitate subsequent recall, at least sometimes (Hogan & Kintsch, 1971, Exp. 1; Runquist, 1983, Exp. 1), and items tested initially in either MC or SA format still show a testing effect when the format is reversed on the final test (Nungester & Duchastel, 1982). However, to confidently adjudicate between the transfer appropriate processing and retrieval effort hypotheses, it is necessary to manipulate the formats of both the intervening and final tests, and only a handful of studies have done so. Support for the transfer appropriate processing view comes from a study by Duchastel and Nungester (1982), who found that taking an intervening MC test produced better performance on a final MC test than taking an intervening SA test, and likewise taking an intervening SA test produced (numerically) better performance on a final SA test than taking an intervening MC test. The authors attributed this benefit to performance for items tested in the same format as before to a "test practice effect".

There has also been some evidence that the more demanding or effortful the retrieval, the greater the enhancement to later memory performance. Glover (1989, Exps 4a, 4b, and 4c) compared the effect of three different types of intervening tests—free recall, cued recall, and recognition—on different types of final tests 4 days later. Regardless of the format of the final test, subjects who received an intervening free recall test performed best, followed in order by those who received an intervening cued recall test, those who received an intervening recognition test, and those who did not receive a prior test. Glover assumed that the amount or completeness of retrieval processing increased from recognition to cued recall to free recall, and concluded that the more complete the retrieval operations during the intervening test, the greater the benefit to final memory performance. Unfortunately, the number of idea units from the studied passage tested on the intervening tests was not equated across the three test types (e.g., on the cued recall test, 12 of the 24 idea units were tested, whereas on the recognition test, six idea units were tested with six distractor sentences), thus the possibility that differential amounts of testing led to the pattern of results cannot be definitively precluded. Also, Glover did not include a condition where subjects were reexposed to the material without taking a test; thus it is possible that it was the re-presentation at test—not necessarily retrieval during the test—that produced the effect.

More recently, Carpenter and DeLosh (2006, Exp. 1), using word lists and including a restudy control condition, replicated Glover (1989, Exp. 4). However, it is remarkable that on all the types of final tests, the intervening free recall condition, which produced the best performance relative to the other intervening test conditions, did not significantly outperform the restudy condition. This could possibly be due to the rather brief delay (i.e., 5 min) before the administration of the final test, as retention interval has been shown to be an important moderator of the testing effect (Roediger & Karpicke, 2006b; Wenger et al., 1980; Wheeler et al., 2003).

The present experiments examined the theoretical underpinnings of the testing effect in a manner that would have direct application to pedagogical practice. Like Glover (1989, Exp. 4) and Carpenter and DeLosh (2006, Exp. 1), we factorially manipulated the formats of the intervening and final tests, with a within-subjects design (Glover used a between-subjects design, and Carpenter & DeLosh used a mixed design). Unlike Carpenter and DeLosh, however, we used educationally relevant journal papers as our study material instead of word lists, and our retention interval before the final test was longer (i.e., 3 days instead of 5 min). Also, unlike Glover, we included a condition in which subjects read equivalent target statements, which provided a focused reexposure to the target material, so as to determine to what extent prior testing boosts later retention beyond re-presentation. If the processes engaged during memory retrieval are crucially responsible for the testing effect, then one might expect that the more demanding or effortful the retrieval during a test, the better that material will be remembered later. This retrieval demands hypothesis would predict that an intervening SA test would result in better performance on the final test than an intervening MC test, regardless of whether the final test was MC or SA format. A straightforward prediction from the transfer appropriate processing framework would be that performance on a final memory test benefits most when the test format matches that of the earlier test, in which case an intervening MC test would enhance final MC items more than an intervening SA test, and an intervening SA test would enhance final SA items more than an intervening MC test. Such a prediction, of course, presumes that MC and SA tests engage disparate memorial operations or processes.

Two experiments were conducted, and they were identical except that the second experiment incorporated feedback. In both experiments, subjects studied brief journal papers and then received either an MC test, an SA test, read statements that repeated the relevant information, or did a filler questionnaire. Three days later, subjects returned for a final test. Experiment 1 was conducted without the provision of feedback to subjects, as an analogue of situations when a classroom instructor dispenses with feedback. In Experiment 2, the correct answer was provided after subjects answered each question on the initial test. This was done to examine the role of corrective feedback in test enhanced learning.

EXPERIMENT 1

Method

Subjects

Forty-eight undergraduates from the Washington University Psychology Subject Pool participated in partial fulfilment of course requirements or for $20 cash.

Materials

Study passages. Four papers from the journal *Current Directions in Psychological Science* (American Psychological Society) were selected as study material. Tables and figures, if present, were removed to homogenise the papers. (Information contained in the tables and figures was redundant for these papers, and hence their removal did not compromise the coherence of the papers.) The average length of the papers was about 2500 words.

Tests. From each paper, eight facts or concepts were selected. These facts were tested in multiple choice (MC) and short answer (SA) formats. In the MC format, subjects had to choose a response among four options, whereas in the SA format, they had to fill in the blank or generate a phrase or sentence to answer the question (questions taken from Anastasio, Rose, & Chapman, 1999; Eagly, Kulesa, Chen, & Chaiken, 2001; Garry & Polaschek, 2000; Treiman, 2000; see Appendix). These facts were also rephrased into one-sentence statements for use in the read statements condition, where subjects read the answers to the test questions without actually attempting the questions.

Design

A 4 (intervening task: MC, SA, read statements, or filler/control) × 2 (final test format: MC or SA) within-subjects design was used. During the first session, subjects studied the four papers. The order of the intervening tasks was kept constant across subjects: They took an MC test immediately after reading the first paper, took an SA test after reading the second paper, read a list of statements (which corresponded to answers to test questions) after reading the third paper, and completed a filler questionnaire after reading the fourth paper. The order of the four papers used was fully counterbalanced across subjects.

During the second session 3 days later, subjects were tested on all four papers. In this final test, questions alternated between MC and SA formats (i.e., each paper was tested with a total of four MC and four SA questions). This final test was administered in two forms, either odd-numbered questions given in MC format and even-numbered questions given in SA format or vice versa, counterbalanced across subjects. The facts tested in this final test were identical to those tested in the first session (for conditions in which subjects took an intervening test), although the question format differed for half of the questions.

An additional 24 subjects were tested on the final tests only (i.e., without having read the four papers), to gain a baseline measure and ensure that performance in all our other conditions was above that baseline.

Procedure

Subjects were tested in groups of 10 or fewer during two experimental sessions. During the first session, subjects were seated at computer terminals, and were given paper copies of the papers, one at a time, and asked to read them carefully because they would be tested on the material later. At the outset, subjects were told to expect different types of tests after each paper, although they never knew the nature of the test prior to reading any specific paper. They were also told to feel free to underline or mark any part of a paper during reading. They were given 15 min to read each paper, and a timer on the computer screen counted down the minutes. After the 15 min elapsed, subjects were instructed to put away the paper. In the conditions in which subjects received a test, test questions appeared on the computer screen successively, one at a time, and subjects wrote their answers on response sheets provided. The test was self-paced, and subjects were told not to amend previous responses once they had advanced to the subsequent questions. After subjects completed the test, they proceeded to read the next paper. In the read statements condition, subjects were given a list of eight statements to read at their own pace after the paper, although a 3-min delay was inserted into the computer program, such that subjects could not proceed to the next paper before 3 min had elapsed. In the control condition, subjects completed a filler questionnaire after reading the paper, after which they were dismissed and reminded to return for the second session. The first session lasted about 1 hour and 20 min.

Three days later (a window period of 70–74 hours after the start of the first session was allowed), subjects returned for the second session. They were tested on all the four papers they had previously read, in both MC and SA formats. As before, test questions appeared one at a time on the computer screen, and subjects wrote their answers on response sheets provided. The second session took about 15 min. At the end of the experiment, subjects were debriefed and thanked for their participation.

Results

Scoring

For MC questions, responses were counted as either correct (1 point) or incorrect (0 points). For SA questions, responses were judged as either correct (1 point), partially correct (½ point), or incorrect (0 points). Scoring was done by a single rater. As a reliability check, 10% of the SA responses were submitted to a second rater for scoring. The interrater agreement and reliability were both .94.

Initial test performance

Although the focus is on final test performance, intervening test performance was also examined. For each participant, we computed the proportion of items correctly answered on the intervening MC ($M = 0.86$, $SD = 0.14$) and SA ($M = 0.54$, $SD = 0.23$) tests. The much higher performance on the MC relative to the SA test has implications for their effects on the later tests, as discussed below.

Final test performance

The proportion of questions answered correctly by each participant from each intervening task condition was computed separately for the two final test formats (MC and SA), and the means can be seen in Figure 1. Due to scaling differences between the MC and SA tests, we analysed the final MC and SA performance separately using one-way repeated measures ANOVAs, with intervening task as a within-subjects factor. Baseline performance of subjects who took the tests without having read the papers was .30 and .02 for the MC and SA tests, respectively, far below performance in all four conditions. The α-level for all analyses was set at .05.

Multiple choice. The type of intervening task did affect final MC performance, $F(3, 141) = 5.54$, $MSE = 0.62$, $\eta^2 = .11$. Post hoc comparisons using paired samples *t*-tests revealed that the intervening MC test condition had greater final performance than the intervening SA test and the control conditions, $t(47) = 2.14$, $d = 0.36$, and $t(47) = 3.08$, $d = 0.61$, respectively. Similarly, the read statements condition had greater final performance than the intervening SA test and the control conditions, $t(47) = 2.05$, $d = 0.40$, and $t(47) = 3.64$, $d = 0.66$, respectively. No other pairwise comparison was significantly different.

Short answer. The type of intervening task did affect final SA performance, $F(3, 141) = 10.85$, $MSE = 4.06$, $\eta^2 = .19$. Post hoc comparisons

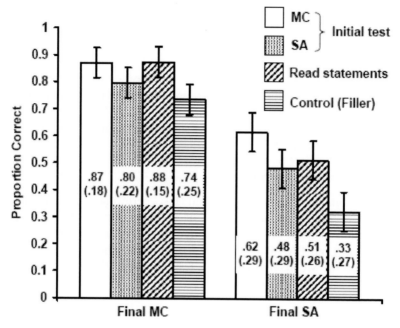

Figure 1. Mean final test performance as a function of intervening task in Experiment 1. Error bars are 95% confidence intervals. *M*s and *SD*s for each condition are listed in the respective bars.

using paired samples *t*-tests indicated that the intervening MC test condition had greater final performance than the intervening SA test and the control conditions, $t(47) = 2.38$, $d = 0.45$, and $t(47) = 6.01$, $d = 1.03$, respectively, and marginally higher final performance than the read statements condition, $t(47) = 1.91$, $p = .06$, $d = 0.37$. The intervening SA test and read statements conditions both had greater final performance than the control condition, $t(47) = 2.91$, $d = 0.55$, and $t(47) = 4.34$, $d = 0.71$, respectively, but were not significantly different from each other.

Lures on MC test.   Prior research has shown that MC tests may cause interference when subjects select lure items and hence acquire false knowledge (Roediger & Marsh, 2005). A supplementary analysis was done to ascertain the proportion of instances in which wrong responses on the intervening MC test led to the same incorrect response being endorsed or produced on the final test. If a subject got an item wrong on the intervening MC test, the probability of endorsing or producing the same incorrect answer was .65 and .11 on the final MC and SA tests, respectively.

Effectiveness score

Due to the different levels of performance on the initial MC and SA tests, we analysed test performance using a metric introduced by Lockhart (1975) and extended by Bjork, Hofacker, and Burns (1981) to assess the degree to which subsequent retrieval is enhanced by prior retrieval, while avoiding item selection artifacts that can compromise the interpretation of raw conditional probabilities. The effectiveness score, a, is based on a simple finite state model that classifies target items into one of four states: whether or not an item is retrievable at the initial test (C or N, respectively) and whether or not it is retrievable at the final test (C or N, respectively). An item in state CN is one that would be retrieved on the initial test (if there was an initial test), but would not be retrieved on the final test. Only items in this state can potentially be facilitated by an act of initial retrieval. The a score is an estimate of the probability that items in state CN make a transition into state CC as a result of an initial retrieval being permitted to occur, and the formula is

$$a = \frac{P_T - P_{NT}}{P_{CN} + P_T - P_{NT}},$$

where P_T = final performance for the conditions in which an intervening test was given, P_{NT} = final performance for the no intervening test control condition, P_{CN} = the expected proportion of items in state CN (i.e., the conditional probability of getting items wrong on the final test, given correct responding on the intervening test). A more in-depth explanation of how the equation was derived can be found in McDaniel, Kowitz, and Dunay (1989).

For the purpose of this analysis, we rescored the responses to SA questions such that responses that were previously scored either as partially or fully correct were now considered correct or retrievable (i.e., C). The a scores could not be calculated in some cases due either to the denominator of the equation being zero, or the conditional probability P_{CN} being undefined (see Table 1 for the means; the number of cases each mean is based on is given in parentheses beside each mean). There was no difference in a scores for either the MC, $F(1, 24) = 0.345$, or SA, $F(1, 34) = 0.01$, final test as a function of intervening test format. This analysis suggests that successful retrieval on an intervening MC test was as effective as successful retrieval on an intervening SA test in enhancing subsequent retrieval, for both final test formats. It should be noted that this conclusion is, at best, tentative due to the rather large number of missing values.

TABLE 1
Mean *a* scores as a function of format of intervening and final test and feedback

Test format		a score		
Intervening test	Final test	Experiment 1–no feedback (no fb)	Experiment 2–feedback (fb)	Difference (fb–no fb)
MC	MC	.86 (n = 35)	.64 (n = 36)	−.22
MC	SA	.38 (n = 45)	.32 (n = 47)	−.06
SA	MC	.79 (n = 30)	.91 (n = 36)	.12
SA	SA	.41 (n = 38)	.67 (n = 40)	.26

The *a* scores could not be calculated in some cases due either to the denominator of the equation being zero, or the conditional probability P_{CN} being undefined. The number of cases that each mean *a* score is based on is given in parentheses. Total $n = 48$.

Discussion

The results in Figure 1 showed that taking an intervening MC test boosted final test performance more than taking an intervening SA test, regardless of whether the final test was MC or SA format. Compared to the control condition, having an intervening MC test resulted in a robust testing effect. Having an intervening SA test led to significantly better performance than the control condition only when the final test was SA format. Different types of tests provide differentially effective cues for accessibility of memories (Tulving & Pearlstone, 1966). Having an intervening SA test may have increased the accessibility to target items, leading to better performance on a subsequent SA test (relative to the control condition), but this effect may not have been apparent on a subsequent MC test because accessibility is already high on a recognition test where copy cues are provided (Darley & Murdock, 1971; Hogan & Kintsch, 1971). The intervening MC test condition was not significantly different from the read statements condition in the final test performance, thus suggesting that the enhancement in retention due to prior MC testing may be equivalent to a focused restudying of the target facts.

One issue that clouds interpretation of the results in Experiment 1 is the differing levels of performance on the initial tests. MC performance (86%) was much higher than SA performance (54%), so it may be no surprise that the greater testing effect shown in Figure 1 for MC tests could have been due to this factor. In addition, in the condition in which subjects restudied the critical facts, they were of course exposed to 100% of them, which may be why even MC testing did not show an advantage relative to reading. Of course, the control of having subjects read the critical statements that would be later tested is quite conservative (and unrealistic) in that students would not normally be able to selectively study only facts on the upcoming test. Reading the entire paper might be a more externally valid rereading control.

Because of these difficulties in comparing MC and SA tests with their varying levels of performance, we used the effectiveness measure (*a* scores). In this analysis, the results showed that SA and MC tests were equally effective in producing gains on the final test when the no test control was used as a baseline. The conclusion differs from that which arises from using the raw scores (Figure 1), but as noted above the analysis should be considered tentative due to large numbers of missing data. Experiment 2 was conducted using the same conditions and design as in Experiment 1, except that feedback was provided after subjects answered the initial test questions so as to equate for exposure to the material on the initial test. Providing feedback in this way should permit a more accurate assessment of how MC and SA testing affects final criterial performance without differences in test performance playing so great a role.

EXPERIMENT 2

The idea that feedback plays an important role in learning is not new (Thorndike, 1913). Kulhavy (1977) proposed that the crucial instructional significance of feedback is to correct erroneous responses during tests. In Experiment 2, subjects were presented with the correct answer after they responded to each question on the intervening tests. The purpose was to ensure that the intervening test conditions would not be penalised by lower exposure to accurate information (relative to the read statements condition), since performance on the initial tests was not at ceiling. Given that supplying the correct answer after subjects respond on an initial test, compared to conditions providing no feedback or feedback that merely states whether or not a response was correct, has been found to greatly augment final retention of verbal material (e.g., Pashler, Cepeda, Wixted, & Rohrer, 2005), we expected the inclusion of feedback on the intervening tests to allow the effect of testing to manifest itself more fully.

Method

Subjects

Fifty-five undergraduates from the Washington University Psychology Subject Pool participated in partial fulfilment of course requirements or for $20 cash. Seven subjects either failed to return for the second session or did not fully adhere to instructions, so their data were excluded from the analysis (leaving data from 48 subjects).

Materials and design

These were the same as in Experiment 1, except that subjects received feedback after each response on the intervening tests in the first session.

Procedure

This was identical to Experiment 1, with the exception that after completing each question on the MC and SA tests in the first session subjects would press a key and the correct answer would appear on the computer screen. They were instructed to press the key only after they had finished responding to each question, and not to change their responses after feedback was provided. Subjects viewed the feedback and proceeded to the following questions at their own pace.

Results and discussion

Initial test performance

Although the focus is on final test performance, the performance on the intervening tests was also examined. For each participant, we computed the proportion of items correctly answered on the intervening MC ($M = 0.85$, $SD = 0.15$) and SA ($M = 0.56$, $SD = 0.19$) tests. Again, MC performance was much higher, but because subjects received feedback immediately, we can assume that initial exposure to correct answers was equated.

Final test performance

The proportion of questions answered correctly by each participant from each intervening task condition was computed separately for the two final test formats (MC and SA), and the means can be seen in Figure 2. Again, the final MC and SA performance was analysed separately using one-way repeated measures ANOVAs, with intervening task as a within-subjects factor.

Multiple choice. There was a significant main effect of intervening task, $F(3, 141) = 14.19$, $MSE = 0.60$, $\eta^2 = .23$. Post hoc comparisons using paired samples t-tests indicated that the intervening SA test condition yielded higher final performance than the intervening MC test, read statements, and control conditions, $t(47) = 2.55$, $d = 0.41$; $t(47) = 3.23$, $d = 0.62$; $t(47) = 6.24$, $d = 1.18$, respectively. The intervening MC test and read statements conditions both had greater final performance than the control condition, $t(47) = 4.22$, $d = 0.78$, and $t(47) = 2.80$, $d = 0.61$, respectively, but were not significantly different from each other.

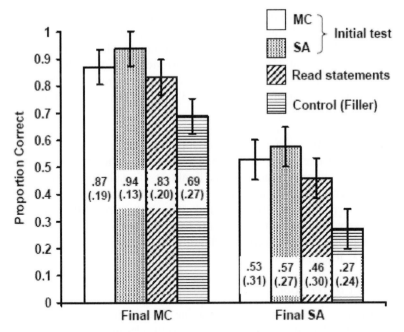

Figure 2. Mean final test performance as a function of intervening task in Experiment 2. Error bars are 95% confidence intervals. *M*s and *SD*s for each condition are listed in the respective bars.

Short answer. There was a significant main effect of intervening task, $F(3, 141) = 12.92$, $MSE = 4.19$, $\eta^2 = .22$. Post hoc comparisons using paired samples *t*-tests revealed that the intervening SA test condition had higher final performance than the read statements and control conditions, $t(47) = 2.18$, $d = 0.40$, and $t(47) = 6.70$, $d = 1.18$, respectively, but performance was not significantly different from the intervening MC test condition. The intervening MC test and read statements conditions both had higher final performance than the control condition, $t(47) = 4.94$, $d = 0.93$, and $t(47) = 3.66$, $d = 0.69$, respectively, but they were not significantly different from each other.

In summary, the results showed that having an intervening SA test enhanced final test performance the most (relative to the read statements or no test control conditions). More importantly, this increase in final retention due to taking an SA test was superior to being presented with statements that reflected answers to test questions (i.e., the read statements condition). Although an intervening MC test did enhance final retention more than the control condition, this effect of testing was not significantly greater than the read statements condition.

The provision of feedback during the intervening tests led to a rather different pattern of results from Experiment 1. The superiority of the intervening SA test condition on both final test formats (MC and SA), relative to the read statements condition, supports the idea that the more retrieval effort expended during a test, the greater the benefit to retention. Even when retrieval is unsuccessful or erroneous on the intervening test, if corrective feedback is supplied, the benefit to final retention from taking a SA test exceeds that of being presented with the test answers for additional study.

Lures on MC test.   Again, a supplementary analysis was done to ascertain the likelihood of a wrong response on the intervening MC test being endorsed or produced on the final test. If a subject got an item wrong on the intervening MC test, the probability of endorsing or producing the same incorrect answer was .30 and .03 on the final MC and SA tests, respectively. Compared to the results from Experiment 1, it is clear that the provision of corrective feedback diminished the negative effect of MC lures (Butler & Roediger, 2006).

Effect of feedback

Final test performance.   To examine directly whether the role of test format in enhancing final retention was moderated by the provision of corrective feedback during the intervening tests, we compared the data from Experiments 1 and 2, and analysed final test performance using a mixed ANOVA with intervening and final test formats as within-subjects factors, and feedback as a between-subjects factor. Crucially, intervening test format interacted with feedback, $F(1, 94) = 11.44$, $MSE = 0.05$, partial $\eta^2 = .11$. Post hoc comparisons using independent samples t-tests revealed that whereas the provision of feedback during the intervening SA tests led to greater final test performance, $t(190) = 2.75$, $d = 0.40$, feedback during the intervening MC tests did not make a difference to final performance, $t(190) = -1.06$, $p = .29$.

Effectiveness score.   An alternative way to look at the role of feedback is to examine its impact on *a*. Again, we compared the data from Experiments 1 and 2, and for each subject, the *a* score was calculated for each intervening (MC and SA) and final (MC and SA) test condition. Since Experiment 2 incorporated corrective feedback during the intervening tests, the *a* scores derived from those data would gauge not only the impact of initial retrieval, but also of feedback as well. As in Experiment 1, the *a* scores could not be calculated in some cases (see Table 1 for the means, with the number of cases contributing to each mean shown in parentheses).

The *a* scores were analysed using a mixed ANOVA with intervening and final test formats as within-subjects factors, and feedback as a between-subjects factor. An intervening SA test yielded higher a scores than an intervening MC test, $F(1, 45) = 7.87$, $MSE = 0.23$, partial $\eta^2 = .15$. But this main effect was qualified by an interaction between intervening test format and feedback, $F(1, 45) = 8.14$, $MSE = 0.23$, partial $\eta^2 = .15$. As can be seen in the last column of Table 1 showing the differences in a scores as a function of feedback, while the provision of feedback did not significantly affect a scores in the intervening MC condition, $t(161) = -1.10$, $p = .27$, feedback marginally boosted a scores in the intervening SA condition, $t(142) = 1.67$, $p < .10$.

Once again, it should be noted that the results of the a score analysis are provisional, given the large number of missing values. Nevertheless, the combined analysis of data from Experiments 1 and 2, both in terms of final test performance and a scores, suggests that SA tests paired with corrective feedback are especially efficacious for enhancing final retention.

The conditional probabilities of correctly recalling an item on the final test given the accuracy of response at the intervening test also point to the same conclusion (see Table 2, which displays the conditional probabilities of being correct on the final test, broken down by final test format, intervening test format, and accuracy at the intervening test). The effect of feedback was more pronounced on items that were unretrievable at the intervening test, so given the lower performance on the intervening SA test (relative to MC test), more items in that condition would benefit from feedback. For items that were answered wrongly on the intervening MC test, with the provision of feedback the proportion answered correctly on the final MC test increased from .27 (Experiment 1) to .52 (Experiment 2), and for the final SA test the increase was from .11 (Experiment 1) to .19 (Experiment 2). For items that were answered wrongly on the intervening SA test, with the provision of

TABLE 2
Conditional probabilities of getting item correct on the final test as a function of test format and feedback

Final test	Intervening test format	Accuracy at intervening test	Experiment 1 (no fb)	Experiment 2 (fb)	Difference (fb−no fb)
MC correct	MC	Wrong	.27	.52	.25
		Correct	.96	.93	−.03
	SA	Wrong	.59	.84	.25
		Partially correct	.90	.94	.04
		Correct	.94	1.00	.06
SA correct	MC	Wrong	.11	.19	.08
		Correct	.62	.51	−.11
	SA	Wrong	.06	.27	.21
		Partially correct	.17	.26	.09
		Correct	.76	.76	.00

feedback the proportion answered correctly on the final MC test increased from .59 (Experiment 1) to .84 (Experiment 2), and for the final SA test the increase was from .06 (Experiment 1) to .27 (Experiment 2). In sum, the conditional probability of getting an item correct on the final test given that the response was wrong or omitted at the intervening test increased with the provision of feedback, and this increase for the intervening SA condition was numerically greater than the intervening MC condition.

GENERAL DISCUSSION

The results of the experiments reported here provide further evidence that testing can improve retention of studied material. More specifically, this study showed that test format and corrective feedback do modulate the testing effect. Taking an SA test was found to boost final test performance more than additional focused exposure to test-relevant information, if corrective feedback was provided to ameliorate poor initial test performance. Although taking an MC test improved final retention relative to doing an unrelated filler task after study, performance was not significantly different from receiving additional exposure to test-relevant information. Based on the current results, taking a MC test may seem functionally equivalent to a reexposure to target information. However, such a conclusion may be premature because in the final SA tests, a tendency existed in both experiments for the intervening MC condition to outperform the read statements condition (and this effect was marginally significant in Experiment 1). So it could be that the benefit of taking an intervening MC test only begins to show when retention is assessed in a manner that is sensitive to the accessibility of target items in memory (Hogan & Kintsch, 1971). Also, it is worth noting again that focused reading of target facts is not the norm in the classroom; it is probably more common for students to reread the entire passage or set of materials. Further research is needed to ascertain if and how the effect of taking an MC test differs from simply reading the material again, but both produced gains relative to the control condition that received no additional exposure to the material.

The pattern of results obtained was not predicted by the transfer appropriate processing framework. A match in the formats of the intervening and final tests did not result in the best final performance. Perhaps a way to reconcile our findings with such a framework is to think of MC and SA tests as not so much engaging fundamentally different processes; rather, MC and SA tests possibly differ in terms of the degree or amount of particular memory processes that are required. In terms of Jacoby's (1991) dual-process model, it could be that SA tests rely more on the intentional, recollective component than MC tests, and that this deeper engagement of recollection during an intervening SA test results in better retention. MC tests, on the other hand, may rely relatively more on familiarity, so an intervening MC test may have less positive transfer to a final SA test (as predicted by the transfer appropriate processing framework). In addition, performance on a final MC test is compromised by the increased familiarity of the incorrect lures seen on the previous MC test (Roediger & Marsh, 2005). Therefore, expecting an intervening MC test to transfer most to a final MC test and an intervening SA test to transfer most to a final SA test may be too simplistic an application of the transfer appropriate processing framework.

Our findings contradict those of Duchastel and Nungester (1982), who found that taking an initial MC test benefited a final MC test more than taking an initial SA test, and taking an initial SA test benefited a final SA test more than taking an initial MC test (although this latter comparison did not reach statistical significance). Possible factors that could have contributed to the discrepancy in findings include differences in the amount of information required for each test question (i.e., their test questions required a brief—often one-word—answer, whereas our test questions often required more elaborate sentence-length answers), differences in retention interval (i.e., their final test was administered after a 2-week delay, whereas our final test was 3 days after the study phase), and the provision of feedback (i.e., they did not provide feedback during the initial test, whereas we did so in Experiment 2). The level of performance on the initial tests could also have played a role; unfortunately, Duchastel and Nungester did not report their initial test results. Also, Duchastel and Nungester conducted their study on students in actual, intact classes, and a questionnaire they administered at the end of the study revealed that 29% of the students discussed the studied passage with their friends during the 2-week retention interval. This could have introduced additional confounds beyond the effect of the initial test.

Other sources of evidence that congruence between the intervening and final tests plays a role in the testing effect come from studies that have shown that when the test questions in the intervening and final tests are phrased similarly, final performance is better than when the questions are paraphrased (Anderson & Biddle, 1975; McDaniel & Fisher, 1991). Also, McDaniel et al. (1989) varied the type of the cues (phonemic or semantic) in a cued-recall test, and found that final recall performance was best when the cues on the final test were of the same type as the cues on the initial test. Future research will be needed to determine the circumstances in which the overlap in the processes engaged by the initial and final tests contributes to the testing effect.

The effect of retrieval

The superiority of taking an intervening SA test with feedback over an MC test on final retention, regardless of the final test format, strongly implicates the role of retrieval effort in improving retention. The findings suggest that the greater the depth or difficulty of the retrieval attempt, the greater the benefit to retention, in a way replicating Glover (1989, Exp. 4) and Carpenter and DeLosh (2006, Exp. 1). Glover and Carpenter and DeLosh did not provide corrective feedback during the initial test or retrieval, but yet found that the most demanding intervening retrieval condition produced the best final performance, whereas in our study the SA test was most effective only when feedback was provided (Experiment 2). The likely explanation is that initial retrieval levels were lower in our study than in theirs, and hence it was necessary for corrective feedback to restore the effectiveness of testing, which is reduced when performance at initial retrieval is low (Wenger et al., 1980).

A recent study by McDaniel, Anderson, Derbish, and Morrisette (2007, this issue), conducted on students enrolled in an online college course, found that on criterial MC exams, those who took weekly SA quizzes out-performed those who took weekly MC quizzes or read target facts (feedback was provided during the weekly quizzes). Their findings dovetail nicely with the results of Experiment 2, and provide additional support for the idea that recall tests are more beneficial for subsequent retention than recognition tests or additional study. Bjork (1975) suggested two possible ways by which retrieval attempts result in the testing effect: The more effortful or difficult the retrieval, (a) the more the memory trace or representation is strengthened, becoming more durable and less vulnerable to interference, and (b) the more the retrieval routes are elaborated and multiplied, increasing accessibility for subsequent retrieval. The findings of the present study are consistent with such an account.

The effect of SA testing bears noticeable similarity to the generation effect. It has been amply demonstrated that when target items are self-generated by subjects in response to cues provided by the experimenter, those items are better retained than items merely presented to be read (Slamecka & Graf, 1978). Importantly, Slamecka and Fevreiski (1983) found that the generation effect is obtained even when subjects fail to correctly generate an item, if the correct response is presented after the failed generation attempt. This, of course, does not mean that the causal mechanisms behind the generation effect and the SA testing advantage obtained in our study are identical. Investigations into the generation effect have shown that factors other than retrieval can contribute to the effect (e.g., allocation of attentional resources at encoding; Schmidt, 1990).

The effect of feedback

The advantage of an intervening SA test only became evident in our study when corrective feedback was provided at test, thus revealing feedback as an important moderator of the effects of testing, especially when performance levels are relatively low during the intervening test. These findings are consistent with those of Pashler et al. (2005), who found that in paired associate learning, supplying the correct answer after an incorrect response during an initial cued recall test produced a great boost to final retention 1 week later. Our analysis of the combined data from Experiments 1 and 2 suggests that the effect of providing feedback on final performance depends on the format of the intervening test. Whereas feedback during an intervening SA test significantly benefited final performance, feedback during an intervening MC test seemed not to make a difference.

One possible reason for the interaction between provision of feedback and intervening test format is that intervening SA test performance was lower (relative to intervening MC performance), hence there were more opportunities for feedback to correct errors and improve final performance. Another possibility is that the retrieval processes engendered by the type of intervening test affected the subsequent processing of feedback, such that taking a SA test, with relatively greater retrieval demands, led to more thorough encoding of feedback than taking an MC test. Although this is admittedly speculative based on the current data, there is evidence that feedback processing can be influenced by metacognitive knowledge (e.g., subjective certainty of whether or not a response is correct) and the accuracy of the response (Kulhavy & Stock, 1989). For instance, Bahrick and Hall (2005) proposed that retrieval failures are informative to subjects, and can spur them to modify their encoding strategies during subsequent presentations of the target material. In addition, Butterfield and Metcalfe (2006) found that subjects focus more attention on corrective feedback when they make an erroneous response with high confidence. Also, Auble and Franks (1978) showed that the more subjects puzzled over seemingly incomprehensible sentences (e.g., "The party stalled because the wire straightened") before they were given a key word that rendered the sentences comprehensible (*cockscrew*), the better their retention of the key word on a later test. This benefit of "effort after meaning" resembles the advantage of SA testing with feedback over just reading the test answers (i.e., the read statements condition) on final performance. To better elucidate the role of feedback, future research should directly examine how the format of a memory test can affect metacognitive judgements and processing of corrective feedback. Nevertheless, regardless of the specific mechanisms involved, the present findings have obvious practical implications.

Practical implications for pedagogy

In recent years, there has been increasing appreciation for the need to base educational practice on scientific evidence (US Department of Education, 2003). Unfortunately, relevant empirical findings from basic psychological research are often not disseminated widely in teacher education programmes (Newcombe, 2002). Although psychological science has had some impact on the teaching of reading (Rayner, Foorman, Perfetti, Pesetsky, & Seidenberg, 2001), pertinent research from other domains of cognitive psychology (e.g., human learning and memory) has not had much influence on actual pedagogical practice (Matlin, 2002). Aside from the typical barriers when trying to communicate across disciplines, part of the problem might be researchers failing to see the potential application of their findings, and educators feeling sceptical about whether laboratory findings generalise to the classroom.

A major impetus for this study was the desire to examine a question about testing that would have straightforward implications for teaching practice, using materials and tests that closely approximate college-level course materials. So as not to be bogged down by the current controversy over high-stakes achievement testing, a change in mindset away from assessing achievement would be useful: Administering tests can be a valuable pedagogical tool to enhance learning (Demptser, 1992; Roediger & Karpicke, 2006b). Our findings clearly support this view by showing that of the two test formats often used by teachers—MC and SA—an SA test is more beneficial for long-term retention than restudying. Also, to obtain the benefit of an SA test, it is necessary to provide corrective feedback, especially when performance on the test is not high. This is a point worth noting, especially since educators at the higher levels (e.g., in college in North America) often do not provide corrective feedback after tests, or at least make it inconvenient for students to view the feedback (e.g., require students to make an appointment to view their test forms), so as to save class time or maintain security of questions in a test bank. The timing of the feedback may also be a factor (Butler & Roediger, 2006; Kulik & Kulik, 1988). Whereas feedback in our study (Experiment 2) was provided after each item, in classrooms it is probably more common for feedback to be delayed for at least a few days until the test scripts have been scored. Future research is required to ascertain whether the current findings generalise to a situation where feedback is delayed. Although MC tests are easier to score and hence more convenient to administer, they tend not to be as effective in improving retention as SA tests. Moreover, a disadvantage of MC or recognition tests is that the presence of (incorrect) lures has the potential to create false knowledge; students may subsequently accept these lures as correct (Butler, Marsh, Goode, & Roediger, 2006; Mandler & Rabinowitz,

1981; Roediger & Marsh, 2005), although feedback does ameliorate the negative effect of MC testing (Butler & Roediger, 2006). Carroll, Campbell-Ratcliffe, Murnane, and Perfect (2007, this issue) demonstrated that retrieval practice during an initial cued recall test could, in some situations, impair memory for untested material on a final test (but see Chan, McDermott, & Roediger, 2006). It should be noted that for conditions in which the effect was present, it was short-lived. Items that underwent retrieval practice, on the other hand, displayed relatively larger and more enduring facilitation on the final test. In addition to the memorial advantages, regular testing has other benefits. It encourages preparation for class, reduces test anxiety, and focuses attention on important course content (Snooks, 2004).

In conclusion, educators have at their disposal a readily available instrument for enhancing learning and retention—SA tests. Instead of giving students handouts summarising key points and facts, a better alternative would be the regular administration of SA quizzes, followed by instructor feedback.

REFERENCES

Anastasio, P. A., Rose, K. C., & Chapman, J. (1999). Can the media create public opinion? A social-identity approach. *Current Directions in Psychological Science, 8*, 152–155.

Anderson, R. C., & Biddle, W. B. (1975). On asking people questions about what they are reading. In G. H. Bower (Ed.), *The psychology of learning and motivation: Advances in research and theory* (Vol. 9 (pp. 89–132)). New York: Academic Press.

Anderson, R. S. (1998). Why talk about different ways to grade? The shift from traditional assessment to alternative assessment. *New Directions for Teaching and Learning, 74*, 5–16.

Auble, P. M., & Franks, J. J. (1978). The effects of effort toward comprehension on recall. *Memory and Cognition, 6*, 20–25.

Bahrick, H. A., & Hall, L. K. (2005). The importance of retrieval failures to long-term retention: A metacognitive explanation of the spacing effect. *Journal of Memory and Language, 52*, 566–577.

Bjork, R. A. (1975). Retrieval as a memory modifier: An interpretation of negative recency and related phenomena. In R. L. Solso (Ed.), *Information processing and cognition: The Loyola symposium* (pp. 123–144). Hillsdale, NJ: Lawrence Erlbaum Associates, Inc.

Bjork, R. A., Hofacker, C., & Burns, M. J. (1981, November). *An "effectiveness-ratio" measure of tests as learning events.* Paper presented at the 22nd annual meeting of the Psychonomic Society, Philadelphia, PA.

Blaxton, T. A. (1989). Investigating dissociations among memory measures: Support for a transfer-appropriate processing framework. *Journal of Experimental Psychology: Learning, Memory, and Cognition, 10*, 3–9.

Butler, A. C., Marsh, E. J., Goode, M. K., & Roediger, H. L. (2006). When additional multiple-choice lures aid versus hinder later memory. *Applied Cognitive Psychology, 20*, 941–956.

Butler, A. C., & Roediger, H. L. (2006, May). *Feedback neutralizes the detrimental effects of multiple-choice testing.* Poster presentation at the 18th annual convention of the Association for Psychological Science, New York.

Butterfield, B., & Metcalfe, J. (2006). The correction of errors committed with high confidence. *Metacognition and Learning, 1*, 69–84.

Carpenter, S. K., & DeLosh, E. L. (2006). Impoverished cue support enhances subsequent retention: Support for the elaborative retrieval explanation of the testing effect. *Memory and Cognition, 34*, 268–276.

Carrier, M., & Pashler, H. (1992). The influence of retrieval on retention. *Memory and Cognition, 20*, 633–642.

Carroll, M., Campbell-Ratcliffe, J., Murnane, H., & Perfect, T. (2007). Retrieval-induced forgetting in educational contexts: Monitoring, expertise, text integration, and test format. *European Journal of Cognitive Psychology, 19*, 580–606.

Chan, J. C. K., McDermott, K. B., & Roediger, H. L. (2006). Retrieval-induced facilitation: Initially nontested material can benefit from prior testing of related material. *Journal of Experimental Psychology: General, 135*, 553–571.

Darley, C. F., & Murdock, B. B., Jr. (1971). Effects of prior free recall testing on final recall and recognition. *Journal of Experimental Psychology, 91*, 66–73.

Dempster, F. N. (1992). Using tests to promote learning: A neglected classroom resource. *Journal of Research and Development in Education, 25*, 213–217.

Dempster, F. N. (1996). Distributing and managing the conditions of encoding and practice. In E. L. Bjork & R. A. Bjork (Eds.), *Memory* (pp. 317–344). San Diego, CA: Academic Press.

Dempster, F. N., & Perkins, P. G. (1993). Revitalizing classroom assessment: Using tests to promote learning. *Journal of Instructional Psychology, 20*, 197–203.

Duchastel, P. C., & Nungester, R. J. (1982). Testing effects measured with alternate test forms. *Journal of Educational Research, 75*, 309–313.

Eagly, A. H., Kulesa, P., Chen, S., & Chaiken, S. (2001). Do attitudes affect memory? Tests of the congeniality hypothesis. *Current Directions in Psychological Science, 10*, 5–9.

Garry, M., & Polaschek, D. L. L. (2000). Imagination and memory. *Current Directions in Psychological Science, 9*, 6–10.

Gates, A. I. (1917). Recitation as a factor in memorizing. *Archives of Psychology*, No. 40, 1–104.

Glover, J. A. (1989). The "testing" phenomenon: Not gone but nearly forgotten. *Journal of Educational Psychology, 81*, 392–399.

Halpern, D. F., & Hakel, M. D. (2002). Learning that lasts a lifetime: Teaching for long-term retention and transfer. *New Directions for Teaching and Learning, 89*, 3–7.

Hanawalt, N. G., & Tarr, A. G. (1961). The effect of recall upon recognition. *Journal of Experimental Psychology, 62*, 361–367.

Hogan, R. M., & Kintsch, W. (1971). Differential effects of study and test trials on long-term recognition and recall. *Journal of Verbal Learning and Verbal Behavior, 10*, 562–567.

Jacoby, L. L. (1991). A process dissociation framework: Separating automatic from intentional uses of memory. *Journal of Memory and Language, 30*, 513–541.

Jones, H. E. (1923). Experimental studies of college teaching: The effect of examination on permanence of learning. *Archives of Psychology*, No. 68, 5–70.

Kulhavy, R. W. (1977). Feedback in written instruction. *Review of Educational Research, 47*, 211–232.

Kulhavy, R. W., & Stock, W. A. (1989). Feedback in written instruction: The place of response certitude. *Educational Psychology Review, 1*, 279–308.

Kulik, J. A., & Kulik, C.-L. C. (1988). Timing of feedback and verbal learning. *Review of Educational Research, 58*, 79–97.

Kuo, T. M., & Hirshman, E. (1996). Investigations of the testing effect. *American Journal of Psychology, 109*, 451–464.

Lachman, R., & Laughery, K. R. (1968). Is a test trial a training trial in free recall learning? *Journal of Experimental Psychology, 76*, 40–50.

LaPorte, R. E., & Voss, J. F. (1975). Retention of prose materials as a function of postacquisition testing. *Journal of Educational Psychology, 67*, 259–266.

Lockhart, R. S. (1975). The facilitation of recognition by recall. *Journal of Verbal Learning and Verbal Behavior, 14*, 253–258.

Mandler, G., & Rabinowitz, J. C. (1981). Appearance and reality: Does a recognition test really improve subsequent recall and recognition? *Journal of Experimental Psychology: Human Learning and Memory, 7*, 79–90.

Matlin, M. W. (2002). Cognitive psychology and college-level pedagogy: Two siblings that rarely communicate. *New Directions for Teaching and Learning, 89*, 87–103.

McDaniel, M. A., Anderson, J. L., Derbish, M. H., & Morrisette, N. (2007). Testing the testing effect in the classroom. *European Journal of Cognitive Psychology, 19*, 494–513.

McDaniel, M. A., & Fisher, R. P. (1991). Tests and test feedback as learning sources. *Contemporary Educational Psychology, 16*, 192–201.

McDaniel, M. A., Kowitz, M. D., & Dunay, P. K. (1989). Altering memory through recall: The effects of cue-guided retrieval processing. *Memory and Cognition, 17*, 423–434.

Morris, C. D., Bransford, J. D., & Franks, J. J. (1977). Levels of processing versus transfer appropriate processing. *Journal of Verbal Learning and Verbal Behavior, 16*, 519–533.

Newcombe, N. S. (2002). Biology is to medicine as psychology is to education: True or false? *New Directions for Teaching and Learning, 89*, 9–18.

Nungester, R. J., & Duchastel, P. C. (1982). Testing versus review: Effects on retention. *Journal of Educational Psychology, 74*, 18–22.

Pashler, H., Cepeda, N. J., Wixted, J. T., & Rohrer, D. (2005). When does feedback facilitate learning of words? *Journal of Experimental Psychology: Learning, Memory, and Cognition, 31*, 3–8.

Rayner, K., Foorman, B. R., Perfetti, C. A., Pesetsky, D., & Seidenberg, M. S. (2001). How psychological science informs the teaching of reading. *Psychological Science in the Public Interest, 2*, 31–74.

Roediger, H. L., & Karpicke, J. D. (2006a). The power of testing memory: Basic research and implications for educational practice. *Perspectives on Psychological Science, 1*, 181–210.

Roediger, H. L., & Karpicke, J. D. (2006b). Test-enhanced learning: Taking memory tests improves long-term retention. *Psychological Science, 17*, 249–255.

Roediger, H. L., & Marsh, E. J. (2005). The positive and negative consequences of multiple-choice testing. *Journal of Experimental Psychology: Learning, Memory, and Cognition, 31*, 1155–1159.

Runquist, W. N. (1983). Some effects of remembering on forgetting. *Memory and Cognition, 11*, 641–650.

Runquist, W. N. (1986). The effect of testing on the forgetting of related and unrelated associates. *Canadian Journal of Psychology, 40*, 65–76.

Runquist, W. N. (1987). Retrieval specificity and the attenuation of forgetting by testing. *Canadian Journal of Psychology, 41*, 84–90.

Schmidt, S. R. (1990). A test of resource-allocation explanations of the generation effect. *Bulletin of the Psychonomic Society, 28*, 93–96.

Slamecka, N. J., & Fevreiski, J. (1983). The generation effect when generation fails. *Journal of Verbal Learning and Verbal Behavior, 22*, 153–163.

Slamecka, N. J., & Graf, P. (1978). The generation effect: Delineation of a phenomenon. *Journal of Experimental Psychology: Human Learning and Memory, 4*, 592–604.

Slamecka, N. J., & Katsaiti, L. T. (1988). Normal forgetting of verbal lists as a function of prior testing. *Journal of Experimental Psychology: Learning, Memory, and Cognition, 14*, 716–727.

Snooks, M. K. (2004). Using practice tests on a regular basis to improve student learning. *New Directions for Teaching and Learning, 100*, 109–113.

Spitzer, H. F. (1939). Studies in retention. *Journal of Educational Psychology, 30*, 641–656.

Thompson, C. P., Wenger, S. K., & Bartling, C. A. (1978). How recall facilitates subsequent recall: A reappraisal. *Journal of Experimental Psychology: Human Learning and Memory, 4*, 210–221.

Thorndike, E. L. (1913). *Educational psychology: Vol. 1. The original nature of man.* New York: Columbia University.

Treiman, R. (2000). The foundations of literacy. *Current Directions in Psychological Science, 9*, 89–92.

Tulving, E. (1967). The effects of presentation and recall of material in free-recall learning. *Journal of Verbal Learning and Verbal Behavior, 6*, 175–184.

Tulving, E., & Pearlstone, Z. (1966). Availability versus accessibility of information in memory for words. *Journal of Verbal Learning and Verbal Behavior, 5*, 381–391.

US Department of Education. (2003). *Identifying and implementing educational practices supported by rigorous evidence: A user friendly guide.* Washington, DC: Author.

Wenger, S. K., Thompson, C. P., & Bartling, C. A. (1980). Recall facilitates subsequent recognition. *Journal of Experimental Psychology: Human Learning and Memory, 6*, 135–144.

Wheeler, M. A., Ewers, M., & Buonanno, J. F. (2003). Different rates of forgetting following study versus test trials. *Memory, 11*, 571–580.

Wheeler, M. A., & Roediger, H. L., III. (1992). Disparate effects of repeated testing: Reconciling Ballard's (1913) and Bartlett's (1932) results. *Psychological Science, 3*, 240–245.

APPENDIX: Test questions used in Experiments 1 and 2

From Garry and Polaschek (2000)

MC	SA	Read Statements
1. Loftus (1993) was the first systematic study to show what? a. Detailed false memories for a whole event could be implanted b. Emotional events tend to be particularly salient and memorable c. Counterfactual thoughts can affect people's judgment of outcomes d. People tend to misremember childhood events	1. Loftus (1993) was the first systematic study to show what?	1. Loftus (1993) was the first systematic study to show that detailed false memories for a whole event could be implanted.
2. According to Sarbin (1998), what strategy do people rely on when they try to remember an event that they do not remember? a. Fabrication of the details b. Exhaustive search of their memory store c. Imagination of the event d. Look for retrieval cues in the environment	2. According to Sarbin (1998), what strategy do people rely on when they try to remember an event that they do not remember?	2. According to Sarbin (1998), people rely on imagination as a strategy when trying to remember an event they do not remember.

Appendix (*Continued*)

MC	SA	Read Statements
3. Subjects become more confident they have experienced a counterfactual event after they imagine the event. This is called a. a false memory b. imagination inflation c. illusory vividness d. confirmatory bias	3. Subjects become more confident they have experienced a counterfactual event after they imagine the event. This is called _____.	3. Subjects become more confident they have experienced a counterfactual event after they imagine the event. This is called imagination inflation.
4. What is source confusion? a. Confusing details from an imagined event with details from an experienced event. b. Forgetting what the source of a memory was. c. When the vividness of a memory is no longer a good indicator of its veracity. d. Misattributing content of a memory to the wrong source.	4. What is source confusion?	4. Source confusion occurs when one misattributes the content of a memory to the wrong source.
5. Two mechanisms have been proposed to account for the boost in confidence of having experienced an imagined counterfactual event. One is source confusion, the other is a. strength of memory trace. b. recollection. c. vividness of memory. d. familiarity	5. Two mechanisms have been proposed to account for the boost in confidence of having experienced an imagined counterfactual event. One is source confusion, the other is_____.	5. Two mechanisms have been proposed to account for the boost in confidence of having experienced an imagined counterfactual event. One is source confusion, the other is familiarity.
6. According to Heaps and Nash (1999), which of the following factors predicts people's tendency to become more confident that they have actually experienced an event after imagining the event? a. Their susceptibility to influence of an authoritative person b. The vividness of their mental imagery c. Their predisposition to hypnotic suggestion d. Their arousal to emotional stimuli	6. State a factor that predicts people's tendency to become more confident that they have actually experienced an event after imagining the event (according to Heaps & Nash, 1999).	6. According to Heaps and Nash (1999), a person's predisposition to hypnotic suggestion predicts one's tendency to become more confident that one has actually experienced an event after imagining the event.
7. One might be tempted to regard the confidence boosting effect of imagining an event as merely the statistical phenomenon of a. regression towards the mean b. restriction of range c. homogeneity of regression d. a spurious correlation	7. One might be tempted to regard the confidence boosting effect of imagining an event as merely the statistical phenomenon of _____.	7. One might be tempted to regard the confidence boosting effect of imagining an event as merely the statistical phenomenon of regression towards the mean.
8. Why do the findings of memory-related effects of repeatedly imagination have clinical implications? a. Because patients might be imagining their disorder/illness b. Because various psychotherapy techniques involve imagining situations and actions c. Because the therapist may find it difficult to distinguish reality from imagination d. Because repeated imagination of events can lead to hallucinations	8. Why do the findings of memory-related effects of repeatedly imagination have clinical implications?	8. The findings of memory-related effects of imagination have clinical implications because various psychotherapy techniques involve imagining situations and actions.

From Anastasio et al. (1999)

MC	SA	Read Statements
1. What is one of the most blatant examples of how the media can induce public opinion? a. Biased news coverage. b. "Live" telecast of events. c. Advertisements. d. Selective censorship of news stories.	1. What is one of the most blatant examples of how the media can induce public opinion?	1. One of the most blatant examples of how the media can induce public opinion is via advertisements.
2. What difference did Archer et al. (1983) find in the way men and women are typically portrayed in news photographs? a. Men are often pictured in job-related roles, whereas women feature more prominently in home-related roles. b. Photographs of men tend to be more close-up compared to that of women. c. The facial expressions of men in photographs tend to be more solemn than that of women. d. Men tend to be photographed alone, whereas photos of women tend to feature them in a group.	2. What difference did Archer et al. (1983) find in the way men and women are typically portrayed in news photographs?	2. Archer et al. (1983) found that in news photographs, men tended to be portrayed more close-up than women.
3. Persons depicted in photographs high in "face-ism" tend to be rated as more _____. a. friendly b. confident c. trustworthy d. intelligent	3. Persons depicted in photographs high in "face-ism" tend to be rated as more _____.	3. Persons depicted in photographs high in "face-ism" tend to be rated as more intelligent.
4. According to Mullen et al. (1986), how was newscaster Peter Jennings different when discussing Ronald Reagan's 1984 campaign compared to when he discussed the campaign of Reagan's political opponent? a. Peter Jennings smiled more when discussing Reagan. b. Peter Jennings was more critical of Reagan. c. Peter Jennings used more hand gestures when discussing Reagan. d. Peter Jennings looked directly at the camera more often when discussing Reagan.	4. According to Mullen et al. (1986), how was newscaster Peter Jennings different when discussing Ronald Reagan's 1984 campaign compared to when he discussed the campaign of Reagan's political opponent?	4. According to Mullen et al. (1986), newscaster Peter Jennings smiled more when discussing Ronald Reagan's 1984 campaign compared to when he discussed the campaign of Reagan's political opponent.
5. According to Gilens (1996), how can the media's portrayal of America's poor affect public perception of poverty? a. The media's overrepresentation of African Americans in poverty can create the perception of more blacks in poverty than there actually are. b. The media's portrayal of poor people as lacking in motivation can lead to less public support for social welfare and public assistance. c. The media's portrayal of people in poverty as being lazy can increase negative attitudes toward poor people. d. The media's underrepresentation of certain groups in their portrayal of poverty can lead to those groups being neglected in social welfare policies.	5. According to Gilens (1996), how can the media's portrayal of America's poor affect public perception of poverty?	5. According to Gilens (1996), the media's overrepresentation of African Americans in poverty can create the perception of more blacks in poverty than there actually are.

Appendix (*Continued*)

MC	SA	Read Statements
6. What is the *hostile media bias*? a. The subtle effects of media portrayal on people's perceptions and opinion. b. The media's influence on hostile and aggressive behaviour. c. The media's reinforcement of negative stereotypes of out-group. d. People on both sides of a controversy perceiving the media as hostile to their group.	6. What is the *hostile madia bias*?	6. *Hostile media bias* refers to the phenomenon where people on both sides of a controversy perceive the media as hostile to their group.
7. Advertising that uses an attractive person to promote a product is relying on the _____ route of persuasion. a. central b. secondary c. peripheral d. fundamental	7. Advertising that uses an attractive person to promote a product is relying on the _____ route of persuasion.	7. Advertising that uses an attractive person to promote a product is relying on the peripheral route of persuasion.
8. A study conducted by the authors (which involved subjects judging guilt/innocence of a fraternity member on charges of vandalism) found that the subject's tendency to side with one's in-group disappeared when_____. a. the subject was exposed to the opinion of an authority figure b. the subject was exposed to evenly mixed opinions of in-group and out-group members c. opinions of others were homogeneous and perfectly correlated with group membership d. the subject was given time to consider all the evidence	8. A study conducted by the authors (which involved subjects judging guilt/innocence of a fraternity member on charges of vandalism) found that the subject's tendency to side with one's in-group disappeared when _____.	8. A study conducted by the authors (which involved subjects judging guilt/innocence of a fraternity member on charges of vandalism) found that the subject's tendency to side with one's in-group disappeared when the subject was exposed to evenly mixed opinions of in-group and out-group members.

From Treiman (2000)

MC	SA	Read Statements
1. What is the alphabetic principle? a. appreciating that many languages, in the written form, use a set of symbols or letters b. appreciating how the letters in printed words relate to how the spoken words sound c. appreciating that there are some rules in how letters can be combined in the spelling of words d. appreciating how the spelling of words can be inconsistent	1. What is the alphabetic principle?	1. The alphabetic principle refers to the appreciation of how the letters in printed words relate to how the spoken words sound.
2. What is a phoneme? a. basic sound unit of a language b. the sound structure of a language c. a syllable d. a cluster of consonants	2. What is a phoneme?	2. A phoneme is the basic sound unit of a language.
3. A syllable can be subdivided into a. consonant clusters b. vowel clusters c. letter segments d. onset and rime	3. A syllable can be subdivided into 2 parts: _____ & _____	3. A syllable can be subdivided into 2 parts: onset and rime.

Appendix (*Continued*)

MC	SA	Read Statements
4. Studies have shown that training in _____ can improve reading and spelling ability in children. a. the names of letters b. analyzing linguistic structure c. phonological awareness d. how to spell their names	4. Studies have shown that training in _____ can improve reading and spelling ability in children.	4. Studies have shown that training in phonological awareness can improve reading and spelling ability in children.
5. What has been the implicit assumption about how children learn letter names and letter sounds? a. they learn them via imitating adult speech b. they learn them unconsciously when listening to adults speak c. they learn them via experimentation with different sounds d. they learn them via rote memorization	5. What has been the implicit assumption about how children learn letter names and letter sounds?	5. The implicit assumption has been that children learn letter names and letter sounds via rote memorization.
6. Recent studies by Treiman et al. have found that an important determinant of knowledge of letter-sounds is a. whether the letter's sound occurs in the name of the letter b. whether the letter is voiced or unvoiced c. the place of articulation of the sound d. the spelling of the child's name	6. Recent studies by Treiman et al. have found that an important determinant of knowledge of letter-sounds is _____.	6. Recent studies by Treiman et al. have found that an important determinant of knowledge of letter-sounds is whether the letter's sound occurs in the name of the letter.
7. Young Joe is more likely to know the _____ of the letter 'j' than Alice or Tom. a. place of articulation b. phoneme c. name d. sound	7. Young Joe is more likely to know the _____ of the letter 'j' than Alice or Tom.	7. Young Joe is more likely to know the name of the letter 'j' than Alice or Tom.
8. There is a widespread view that young children are purely _____ readers, memorizing associations between whole printed words and their spoken form	8. There is a widespread view that young children are purely _____ readers, memorizing associations between whole printed words and their spoken form.	8. There is a widespread view that young children are purely logographic readers, memorizing associations between whole printed words and their spoken form.

From Eagly et al. (2001)

MC	SA	Read Statements
1. What is the congeniality hypothesis? a. People are motivated to avoid information that challenges their attitudes. b. People's memories are biased in favor of information that agrees with their attitudes. c. People selectively pay attention only to attitudinally agreeable information. d. People tend to more elaborately process information that is inconsistent with their attitudes.	1. What is the congeniality hypothesis?	1. The congeniality hypothesis proposes that people's memories are biased in favor of information that agrees with their attidues.

Appendix (*Continued*)

MC	*SA*	*Read Statements*
2. In the author's meta-analysis of research on memory for attitude-relevant information, what was the trend in results for later compared to early findings? a. Early experiments showed that congenial information was less memorable than uncongenial information, whereas later research tended to yield the reverse pattern or null difference. b. The results tended to be inconsistent regardless of whether the studies were early or more recent. c. More recent studies tended to yield a larger effect of congeniality on memory than early studies. d. Early experiments showed that congenial information was more memorable than uncongenial information, whereas later research tended to yield the reverse pattern or null difference.	2. In the auhors' meta-analysis of research on memory for attitude-relevant information, what was the trend in results for later compared to early findings?	2. In the auhors' meta-analysis of research on memory for attitude-relevant information, they found early experiments tended to show that congenial information was more memorable than uncongenial information, whereas later research tended to yield the reverse pattern or null difference.
3. What is the likely cause of the different trend in findings for earlier vs. later studies? a. Improvements in the procedures used to assess memory. b. Participants in later studies tended to have less polarized attitudes. c. Participants in earlier studies tended to have weaker attitudes. d. Later studies examined more variables than earlier studies.	3. What is the likely cause of the different trend in findings for earlier vs. later studies?	3. Improvements in the procedures used to assess memory is the likely cause of the different trend in findings for earlier vs. later studies.
4. What is the design/procedure for a typical experiment looking at the congeniality effect? a. Participants attitudes toward an issue are measured before and after presentation of information relevant to the issue. b. Participants are presented with information that disagrees with their attitudes, and their subsequent memory for that information assessed. c. Participants with opposing attitudes toward an issue are presented with information on one or both sides of the issue, and their subsequent memory for that information assessed. d. Participants are presented with information that agrees with their attitudes, and their subsequent memory for that information assessed.	4. What is the design/procedure for a typical experiment looking at the congeniality effect?	4. An experiment looking at the congeniality effect typically has the following design/procedure: Participants with opposing attitudes toward an issue are presented with information on one or both sides of the issue, and their subsequent memory for that information assessed.
5. What did the authors propose could account for the weakness of the congeniality effect shown in experiments that were methodologically more rigorous? a. Attitudes have little impact on memory. b. People avoid information that challenges their attitudes. c. People may mount an active defense and hence thoroughly process counterattitudinal information. d. Participants had insufficiently strong attitudes.	5. What did the authors propose could account for the weakness of the congeniality effect shown in experiments that were methodologically more rigorous?	5. The authors proposed that people may mount an active defense and hence thoroughly process counterattitudinal information, thus accounting for the weakness of the congeniality effect found in experiments that were methodologically more rigorous.

Appendix (*Continued*)

MC	SA	Read Statements
6. In their recent experiment (Eagly et al., 2000), the authors found what difference between congenial and uncongenial information? a. Congenial information was recalled better than uncongenial information. b. Participants had more prior knowledge of congenial information than uncongenial information. c. Uncongenial information was better recalled soon after the message was presented, whereas congenial information was better recalled after a delay. d. Uncongenial information elicited more thought and attention than congenial information.	6. In their recent experiment (Eagly et al., 2000), the authors found what difference between congenial and uncongenial information?	6. In their recent experiment (Eagly et al., 2000), the authors found that uncongenial information elicited more thought and attention than congenial information.
7. The authors propose that to persuade people to accept a position that is highly divergent from their own attitudes, it might be best to a. use an incremental approach whereby each exposure to uncongenial information produces only a small amount of change. b. expose them to large amounts of uncongenial information at one go. c. employ an authority figure to promote the counterattitudinal position. d. encourage them to think global thoughts concerning the issue rather than differentiated thoughts.	7. The authors propose that to persuade people to accept a position that is highly divergent from their own attitudes, it might be best to _____.	7. The authors propose that to persuade people to accept a position that is highly divergent from their own attitudes, it might be best to use an incremental approach whereby each exposure to uncongenial infor-mation produces only a small amount of change.
8. According to dual-process theories of social judgement, a recipient who is lacking in motivation and capacity will likely adopt what type of approach when faced with uncongenial information? a. Yield and capitulate to the counterattitudinal viewpoint. b. Adopt an active resistance and confront the uncongenial information. c. Adopt a passive, avoidant approach and process the information less. d. React emotionally and dismiss the information outright.	8. According to dual-process theories of social judgement, a recipient who is lacking in motivation and capacity will likely adopt what type of approach when faced with uncongenial information?	8. According to dual-process theories if social judgment, a recipient who is lacking in motivation and capacity will likely adopt a passive, avoidant approach and process the information less.

EUROPEAN JOURNAL OF COGNITIVE PSYCHOLOGY
2007, 19 (4/5), 559–579

Improving students' self-evaluation of learning for key concepts in textbook materials

Katherine A. Rawson and John Dunlosky
Kent State University, Kent, OH, USA

Why do students have difficulties judging the correctness of information they recall (e.g., definitions of key concepts in textbooks), and how can students improve their judgement accuracy? To answer these questions, we had college students read six expository passages, each including four key terms with definitions. After reading a text, each key term was presented, and participants (a) attempted to recall the corresponding definition and (b) self-scored the correctness of the response (incorrect, partially correct, or entirely correct). Participants were overconfident, with inflated judgements for responses that were objectively incorrect. When participants could inspect correct definitions while judging their responses, judgement accuracy improved. Counterintuitively, however, some overconfidence remained. We discuss implications of these results for theory, education, and the two questions posed above.

In most educational contexts, students are often presented with large amounts of new material to learn. In primary and secondary grades, current mandates require that students demonstrate competence across many content areas before advancing in school; at the college level, students must learn information from textbooks, lectures, activities, and assignments from several different courses at a time. Given the amount of material that students are expected to learn coupled with a finite amount of time available for study, the task of a successful student is not just to achieve a high level of learning but to regulate study as efficiently as possible.

To regulate study efficiently, a student can profit from accurately evaluating the extent to which they have learned information while studying (Thiede, 1999; Thiede, Anderson, & Therriault, 2003). Consider a student who is studying several different sections in a textbook for an upcoming exam. If the student cannot accurately evaluate how well she has learned a

Correspondence should be addressed to Katherine Rawson, Department of Psychology, Kent State University, PO Box 5190, Kent, OH 44242-0001 USA. E-mail: krawson1@kent.edu

Preparation of this report was supported by Cognition and Student Learning grant No. R305H050038 from the US Department of Education.

DOI: 10.1080/09541440701326022

given section, the efficiency of study may be compromised. For example, the student may decide she has learned information well when in fact she has not. If, as a result, she prematurely terminates study of that information, she obviously will not achieve an adequate level of learning. Alternatively, the student may decide that she has not learned information well when in fact she has. If she unnecessarily prolongs study of that information, she will limit the amount of time available for studying other information that has not yet been well learned.

Thus, efficient regulation of learning can depend in part on how accurately an individual can evaluate his or her own learning of text materials. Unfortunately, previous research has shown that students' evaluations of their own learning for text materials are only moderately accurate at best. The goal of the present research was to evaluate two hypotheses for why the accuracy of students' evaluations of learning is constrained. In so doing, we will also explore a technique that may improve students' evaluations of learning and ultimately support more efficient regulation of learning. In the next section, we offer a brief historical review of the previous research that led to the two hypotheses to be tested here, and we then describe each of these hypotheses in some detail.

WHAT CONSTRAINS THE ACCURACY OF EVALUATIONS OF LEARNING?

A standard method used in much of the previous research involves an analogue to the study context described for our hypothetical student above. Specifically, individuals are presented with several short texts to study and are then asked to make a judgement for each one. Currently, the most common judgement is what we will refer to as a *global prediction*, in which the individual is asked to predict how well they will do on an upcoming test of the material for each text. We refer to these as *global* because only one prediction is made for an entire text. The accuracy of these predictions is then measured by correlating an individual's predicted test performance with actual test performance across texts. With few exceptions, 20 years of research has shown that the accuracy of global predictions is quite poor, with the mean across individual correlations usually around $+.25$ (e.g., Maki, 1998; Miesner & Maki, 2007 this issue; Weaver, Bryant, & Burns, 1995; for recent reports of enhanced accuracy, see Dunlosky & Rawson, 2005; Thiede et al., 2003).

In recent research, Dunlosky, Rawson, and Middleton (2005) proposed that global predictions may be relatively inaccurate because of the mismatch between the grain size of the information a student must consider when evaluating learning (i.e., the entire text) and the grain size of the information being tested (i.e., specific definitions, main points, or key concepts from the text). To minimise this mismatch, another kind of judgement has recently been introduced to the standard method, which we refer to here as a *term-specific prediction*. More specifically, Dunlosky, Rawson, and McDonald (2002) had individuals study six short passages, each containing four key terms (e.g., a passage from a nutrition textbook on the body's energy use, which included definitions and explanations of basal metabolism, thermic effects of food, adaptive thermogenesis, and direct calorimetry). After studying a passage, individuals first made a global prediction. They were then presented with each of the four key terms (e.g., basal metabolism) one at a time for a term-specific prediction, in which they were asked to predict how well they would be able to recall the definition of that term on the subsequent test. They were then tested for their memory of the definition for each term. The accuracy of term-specific predictions was measured by correlating an individual's predicted recall with actual recall across all the terms.

The term-specific predictions were expected to be highly accurate, based on the following three assumptions: (a) When presented with the key term for a term-specific prediction, students would self-test using the term as a cue to try to retrieve the target definition, (b) the outcome of this retrieval attempt would be highly diagnostic of subsequent test performance, and (c) students would base their predictions on the outcome of the retrieval attempt. In contrast to our expectation, however, term-specific accuracy was only moderate (mean across individual correlations = .50) and was not significantly greater than the accuracy of the global predictions (mean across individual correlations = .40).

Subsequent work evaluated the three assumptions that led to the (incorrect) expectation that term-specific predictions would be highly accurate. Pertaining to the first assumption, were students self-testing at the time of the term-specific prediction? To answer this question, Dunlosky et al. (2005) used the same method as described above, with one important modification: Prior to making their term-specific prediction, half of the participants were presented with the key term and were explicitly required to overtly recall the target definition (as in Nelson, Narens, & Dunlosky, 2004; Son & Metcalfe, 2005). Students who were forced to attempt recall of target definitions made more accurate term-specific predictions than students who did not overtly attempt recall ($M = 0.73$ vs. $M = 0.57$). Thus, one reason why the accuracy of term-specific predictions can be constrained is that students do not always spontaneously self-test in order to evaluate their learning (cf. Kelemen, 2000).

Note, however, that the accuracy of term-specific predictions was still less than perfect even when students were forced to attempt recall of the target definitions. With respect to the second assumption, was accuracy constrained because the outcome of the retrieval attempt was not highly diagnostic of subsequent recall? Follow-up analyses showed that the mean intraindividual correlation between prejudgement recall and criterion recall across terms was around .90, suggesting that the outcome of prejudgement recall was a highly diagnostic cue for predicting criterion recall. Why then was the accuracy of term-specific predictions still constrained? Consider again the third assumption, which states that students base their predictions on the outcome of the retrieval attempt. The mean intraindividual correlation between prejudgement recall and term-specific predictions was around .67. Thus, although the outcome of prejudgement recall was diagnostic of subsequent test performance (.90), students apparently did not fully capitalise on this cue when predicting test performance.

Most important for present purposes, a follow-up experiment indicated that students were not adequately evaluating the correctness of the outcome of their retrieval attempts. In this experiment, all students performed prejudgement recall. Importantly, some participants were then asked to make a *self-score judgement*. For this judgement, they were explicitly asked to score the correctness of their recall response, using the following prompt: "If the correctness of the definition you just wrote was being graded, do you think you would receive no credit, partial credit, or full credit?" In evaluating their performance, students often assigned full credit to responses that were only partially correct, partial credit to responses that were completely correct, and most troublesome, they frequently assigned partial or even full credit to responses that were completely incorrect. The present research was designed to explore the inaccuracy of these self-score judgements and the difficulties students have in evaluating the correctness of their own responses when self-testing memory for target information.

WHY ARE SELF-SCORE JUDGEMENTS INACCURATE?

As argued above, students must be able to identify material that they have not learned well enough so that subsequent study can be focused on that content. One strategy for assessing how well something has been learned is to self-test—indeed, the original expectation was that term-specific predictions would be quite accurate because they afforded an opportunity to self-test memory for information at an appropriate grain size for evaluation. However, the value of self-testing for guiding subsequent regulation of study hinges critically on the extent to which an individual can accurately evaluate the *correctness* of the outcome of the self-test, which leads us to our focal question: What factors constrain the accuracy of students' self-evaluations of target information recalled during self-testing of memory for expository text content?

According to the *absence of standard hypothesis*, when students do not have access to the objectively correct target information—i.e., the objective standard of evaluation—they will have difficulties in evaluating the correctness of recalled information. Such difficulty seems inevitable, because if a student retrieves incorrect information, he or she presumably has not adequately learned the actual sought-after target information, so comparing the retrieved information to this sought-after information would be impossible without external support (for an elaboration of this idea, see Hacker, 1998). Without access to an external standard, students may turn to other cues to evaluate the adequacy of the retrieved information, such as the quantity of information recalled during the retrieval attempt, which may not be highly related to the objective correctness of the retrieved information (e.g., Baker & Dunlosky, 2006; Morris, 1990). According to this hypothesis, however, students may be capable of better evaluating the correctness of their recall output if they have access to an external standard. Examining the degree of consistency between a retrieved response and the objectively correct response may serve as a means to more accurately evaluate the correctness of the retrieved response.

The *limited competence hypothesis* states that students have limited ability to evaluate the correctness of retrieved information, even if an external standard is available for comparison. According to this hypothesis, providing students with an external standard for comparison to their generated responses will not significantly improve the accuracy of self-score judgements. The plausibility of this account is suggested by several studies in the text comprehension literature. Research on error detection has shown that students are often unable to identify factual inconsistency between pieces of information within a text. For example, Otero and Kintsch (1992) presented readers with short texts, some of which contained two sentences that were inconsistent with one another (e.g., "Superconductivity ... has only been obtained by cooling certain materials to low temperatures near absolute zero", and then later, "[u]ntil now superconductivity has been achieved by considerably increasing the temperature of certain materials"). Forty per cent of the inconsistencies went undetected. Similarly, Johnson and Seifert (1994) presented readers with fictional police reports about a warehouse fire, in which one of the earlier reports was subsequently updated (e.g., "... they have reports that cans of oil paint and pressurised gas cylinders had been present in the closet before the fire", and later, "... the closet reportedly containing cans of oil paint and gas cylinders had actually been empty before the fire"). When subsequently tested for understanding of the cause of the fire, over 90% of the participants made at least one direct, unqualified reference to the volatile materials. Studies such as these suggest that students may have difficulty recognising the inconsistency between two pieces of explicitly stated text information. Even more relevant, other research suggests that students have difficulty recognising inconsistencies between generated responses and explicit text information. Howe (1970) had readers listen to a short passage, recall it, and then listen to it again. In each of the following 3 weeks, they first recalled the text and then listened to it again. Information that was incorrectly recalled on initial tests continued to be recalled on subsequent tests despite re-presentation of the text (in fact, inclusion of previously recalled incorrect information was two to three times more likely than inclusion of previously unrecalled correct information). This finding is suggestive of individuals' inability to recognise the inconsistency between what they recall and explicitly presented information that is objectively correct.

GOALS OF THE PRESENT RESEARCH

The present research was designed to evaluate the absence of standard hypothesis and the limited competence hypothesis. For this purpose, we adapted the method used by Dunlosky et al. (2005) described above. Participants were presented with several short texts each containing four key terms. After reading a text, readers were presented with each of the four key terms one at a time and were asked to recall the definition of the term. After the recall attempt, all participants were asked to make a self-score judgement in which they rated the correctness of the generated response. For one group of participants, the correct definition was presented along with the participant's generated response at the time of the self-score judgement (hereafter referred to as the *standard* group). For the other group, only the generated response was presented (hereafter referred to as the *no standard* group).

The absence of standard hypothesis predicts that individuals who are presented with the correct answer will make more accurate self-score judgements than individuals who are not shown the correct answer. Specifically, individuals in the standard group will be more likely to assign generated responses to the appropriate categories ("no credit" for incorrect responses, "partial credit" for partially correct responses, and "full credit" for correct responses). In contrast, the limited competence hypothesis predicts that the accuracy of the self-score judgements will not significantly differ for the standard and no standard groups. Note, of course, that these two hypotheses are not mutually exclusive, and both mechanisms may undermine the accuracy of people's self-assessments. Thus, providing an external standard may improve accuracy, yet students' self-score judgements may still demonstrate some biases even with a standard. Importantly, this possibility can be explored using the methods adopted in the present research.

METHOD

Participants and design

Fifty-six undergraduates participated to partially satisfy a course requirement in Introductory Psychology. Participants were randomly assigned to
one of two groups, standard ($n = 30$) or no standard ($n = 26$).

Materials

The materials were the same as those used by Dunlosky et al. (2005) and
included seven expository texts (one sample and six critical) that were taken
from introductory-level textbooks from various undergraduate courses (e.g.,
nutrition, family studies, communication). Texts were between 271 and 281
words long, with Flesch-Kincaid scores ranging from grade levels 10 to 12.
Each text contained four key terms (presented in capital letters), and
each term was immediately followed by a one-sentence definition (e.g.,
"ADAPTIVE THERMOGENESIS refers to when the body expends energy
to produce heat in response to a cold environment or as a result of
overfeeding"). Macintosh computers presented all materials and recorded all
responses.

Procedure

Participants were given detailed instructions about each phase of the task.
Before beginning the critical study trials, participants practised each task
with the sample text and test questions to familiarise them with the kind of
text and tests they would receive in the critical trials.

The critical texts were presented in random order for each participant.
Each text was presented individually in its entirety for self-paced study.
Participants clicked a button on the screen to indicate when they were done
studying a text. Immediately after reading a given text, participants were
asked to make a global prediction with the following prompt: "How well will
you be able to complete a test over this material? 0 = definitely won't be able,
20 = 20% sure I will be able, 40 = 40% sure ... 100 = definitely will be able."
After making the global prediction, participants were presented with the
following prompt: "Please practice recalling the following information from
the text you just read", followed by one of the key terms from the text (e.g.,
"Define: adaptive thermogenesis"). Participants typed their response into a
text field on the screen. After participants indicated they were done recalling
the definition by clicking on a button, they were asked to evaluate or "self-
score" the response they generated. For participants in both groups, the

participant's response appeared on the screen along with the following prompt: "If the correctness of the definition you just wrote was being graded, do you think you would receive no credit, partial credit, or full credit?" For participants in the standard group, the correct definition of the term was also presented above the participant's response. After self-scoring their response, the response (and definition) were removed from the screen. Participants were then asked to predict how well they would be able to define that term on the actual test, using the 0–100 scale described above.

Participants completed this procedure for each of the four terms in a text (prejudgement recall, self-score judgement, and term-specific prediction), and then the criterion test was administered for that text. For the criterion test, each of the four terms was presented individually on the screen, and the participants typed their recall of the definition into a text field. After the criterion test for a text had been completed, participants studied the next text and so on until the procedure had been completed for all six texts.

RESULTS

Self-score judgements

For each participant, we computed a mean across self-score judgements for all items (assigning a value of 0 to a self-score judgement of "no credit", 50 to a judgement of "partial credit", and 100 to a judgement of "full credit"). We then computed a mean across the individual means in each group. Overall, self-score judgements were significantly lower for the standard group ($M = 34$, $SE = 3$) than for the no standard group ($M = 44$, $SE = 3$), $t(54) = 2.25$, $p < .05$.[1]

Of greater interest is examination of the self-score judgements for each kind of prejudgement recall response. Each prejudgement recall response was scored using a gist criterion. Specifically, each definition consisted of two to four idea units, and an idea unit was counted as being present in a participant's response when the participant either stated the idea verbatim or correctly paraphrased the original text. On the basis of this scoring, we then assigned each prejudgement recall response to one of five categories, which were chosen because they reflect the kinds of qualitative outcomes that should be reflected in people's self-score judgements (see Dunlosky et al., 2005). These categories were: omission error (no response), commission error (a completely incorrect response), partially correct (a response

---

[1] For all analyses involving self-score judgements, we also conducted nonparametric inferential tests (Mann-Whitey Test) to compare self-score judgements from the standard group versus the no standard group. All of these tests supported the same statistical conclusions as obtained from the parametric tests.

containing at least one correct idea unit), partial plus commission (a response that contained incorrect information along with at least one correct idea unit), and correct (a completely accurate response). Partial plus commission responses were rare (on average, less than one per participant), so we do not consider this response category further. For each group, the percentage of total responses in each category is indicated above each bar in Figure 1.

For each individual, we computed the mean self-score judgement for responses within each of the four categories of interest. The means across these individual values in each response category for each group are depicted by the bars in Figure 1. A 2 (group) × 4 (response category) mixed factor analysis of variance (ANOVA) revealed significant main effects of group and response category, $F(1, 41) = 9.04$, $MSE = 450.67$, $p < .01$, and $F(3, 123) = 176.66$, $MSE = 253.76$, $p < .001$, respectively, and a significant interaction, $F(3, 123) = 12.10$, $MSE = 253.76$, $p < .001$. Follow-up tests revealed that the magnitude of self-score judgements for the two groups did not significantly differ for omissions, partially correct responses, or correct responses, all $ts \leq 1.20$, $ps > .10$. In contrast, self-score judgements for commissions were significantly lower in the standard group than in the no standard group, $t(54) = 5.92$, $p < .001$. Thus, when individuals were provided standards for evaluating their responses, they were better able to judge the correctness of those responses, consistent with the absence of standard hypothesis. This improvement in the absolute accuracy of the self-score judgements for commissions may be particularly important, considering that commissions were the most frequent kind of response—across participants, 38% of all prejudgement recall responses were commissions (compared to 29% omissions, 16% partially correct responses, and 14% correct responses).

Note, however, that although providing standards improved the accuracy of self-score judgements for commissions, students still inappropriately awarded credit to some responses that were completely incorrect, which provides some support for the limited competence hypothesis. To examine this pattern of errors further, for each individual, we computed the percentage of commissions that were assigned to each of the three self-score judgements (no credit, partial credit, or full credit). We then computed the mean percentage across participants for each judgement category in each group. These means are reported in Table 1. Surprisingly, even when a standard for evaluation was explicitly provided, students still judged incorrect responses as partially or fully correct 43% of the time.

Why are students sometimes unable to recognise the inconsistency between generated responses and explicit standards for evaluation? One possibility is that self-score judgements may be partly influenced by the length of the response. For instance, students may believe that if they

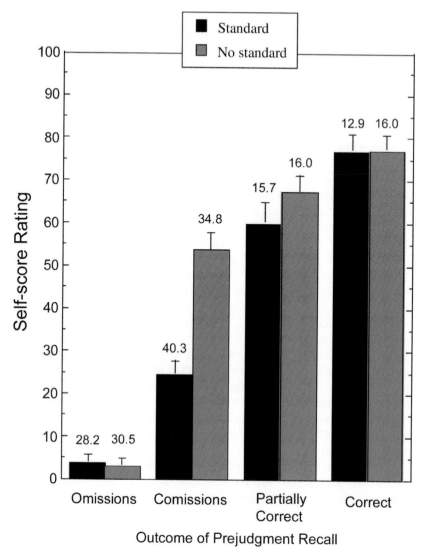

Figure 1.   Bars represent mean self-score judgement magnitude for each of four kinds of response as a function of group. Error bars are standard errors of the mean. Percentages above each bar indicate the percentage of responses (out of 24) assigned to each response category depicted in graph (values do not total to 100 because one low-frequency response category was omitted from analysis; see text for details).

produce enough, something in what they produce must be right. This possibility is consistent with the accessibility hypothesis of metacognitive judgements, which states that judgements reflect the amount and speed with

TABLE 1
Percentage of commissions assigned to each of three self-score judgment
categories

	No credit		Partial credit		Full credit	
	M	SE	M	SE	M	SE
Standard	57	5	37	5	6	2
No standard	17	4	58	4	25	5

SE = standard error of the mean.

which to-be-evaluated information is accessed from memory (e.g., Benjamin, Bjork, & Schwartz, 1998; Koriat, 1993; Morris, 1990). To explore this possibility, for the terms that elicited commission errors, we correlated the number of words in each generated response with the self-score judgements. Whereas response length was significantly correlated with self-score judgements in the no standard group ($r = .29$, $p < .01$), response length was not significantly correlated with self-score judgements in the standard group ($r = .10$, $p = .09$). Thus, the possible effects of response quantity did not strongly bias students' judgements, especially for those who had access to the objective standards while evaluating the correctness of their recall.

Another possibility is that although an explicit standard was presented on every trial, students may not always have referred to the standard when making their self-score judgement. One indicator of whether they referred to the standard can be inferred from how much time participants took to make the self-score judgements (referred to hereafter as *self-score time*). For each individual in each group, we computed (a) the mean self-score time for commission errors that were assigned a self-score judgement of no credit and (b) the mean self-score time for commission errors that were assigned a self-score judgement of either partial credit or full credit. (We combined the latter two judgement categories to reduce the number of participants who would have been dropped from analysis because they did not have any commission errors in one of the categories.) Consider first the self-score times for participants in the no standard group, as these results provide a baseline for when participants could not consult an external standard. Across participants in the no standard group, mean self-score time was 2.5 s ($SE = 0.6$) for commission errors that were assigned no credit and 2.3 s ($SE = 0.2$) for commission errors that were assigned partial or full credit. By comparison, across participants in the standard group, mean self-score time was 7.9 s ($SE = 0.7$) for commission errors that were assigned no credit and 6.9 s ($SE = 0.8$) for commission errors that were assigned partial or full credit. Self-score times were consistently longer in the standard group than in the no standard group, $ts > 4.90$, suggesting that participants in the standard group were consulting the explicit standard when making their self-score judgements. More important, self-score judgement times in the standard group did not significantly differ for commissions assigned no credit and commissions assigned partial or full credit, $t(21) = 1.04$. Although tentative, these results provide further support for the limited competence hypothesis because students were apparently consulting the external standard but still failed to recognise inconsistencies between recalled information and objectively correct information some of the time.

Cued recall performance on the prejudgement test and the criterion test

For prejudgement recall and for criterion test recall, each cued recall response was scored using a gist criterion. Each definition consisted of two to four idea units, and an idea unit was counted as being present in a response if the idea unit was stated verbatim or if it was a correct paraphrase of the original text. A response was then assigned a score of 0% if it contained no correct idea units, 50% if it contained at least 50% but less than 100% of the idea units contained in the definition, and 100% if it contained all of the idea units in the definition. For each participant, we then computed mean recall performance across all individual response scores on prejudgement recall and mean recall performance across all individual response scores on the criterion test; thus, individual mean values could range from 0 to 100%. Mean recall across participants in each group for prejudgement recall and the criterion test are reported in Table 2.

A 2 (group) × 2 (time of test: prejudgement recall vs. criterion test) mixed factor ANOVA yielded a significant main effect of time of test and a significant interaction, $F(1, 54) = 59.46$, $MSE = 58.76$, $p < .001$, and $F(1, 54) = 41.50$, $p < .001$, respectively. The main effect of group was not significant, $F < 1.85$. Follow-up tests revealed that whereas performance significantly improved from prejudgement recall to criterion test for the standard group, performance did not significantly improve in the no standard

TABLE 2
Performance on pre-judgment recall and criterion recall

| | Standard | | No standard | |
	M	SE	M	SE
Prejudgment recall	20	3	24	3
Criterion recall	41	4	25	4

Performance is reported as a percentage. SE = standard error of the mean.

TABLE 3
Performance on criterion recall test as a function of outcome on the prejudgment recall test

Outcome of prejudgment recall	Standard		No standard	
	M	*SE*	*M*	*SE*
Omission	16.5	3.5	2.3	1.0
Commission	35.6	4.8	8.8	3.0
Partially correct	58.2	5.0	35.8	4.2
Correct	91.8	2.7	84.7	5.6

Performance is reported as a percentage. *SE* = standard error of the mean.

group, $t(29) = 8.22$, $p < .001$, and $t(25) = 1.56$, respectively. Moreover, whereas prejudgement recall in the two groups did not significantly differ, criterion test performance was significantly greater in the standard group than in the no standard group, $t(54) = 0.88$ and $t(54) = 3.00$, $p < .01$, respectively.

To further explore the effects of providing an external standard on criterion test performance, we examined criterion test performance as a function of prejudgement recall status. Specifically, for each individual, we computed mean criterion recall conditionalised on the outcomes of prejudgement recall, namely, for omission errors, commission errors, partially correct responses, and correct responses. Means across individual values for each group are reported in Table 3. Criterion recall was greater in the standard group than in the no standard group for prejudgement omissions, commissions, and partially correct responses, $ts > 3.36$. Criterion recall for prejudgement correct responses in the two groups did not significantly differ, $t(46) = 1.16$.

Within the standard group, recall significantly improved from prejudgement to criterion test for omissions, commissions, and partially correct responses, $ts > 3.45$, whereas recall significantly declined from prejudgement to criterion test for correct responses, $t(23) = 2.88$, $p < .01$.[2] Within the no

---

[2] By definition, mean prejudgement recall was 0% for omissions and commissions, and 100% for correct responses. Mean prejudgement recall across partially correct responses was 40% ($SE = 2$) in the standard group and 36% ($SE = 3$) in the no standard group. Note that prejudgement recall for partially correct responses is less than 50% because of the differences between the criterion used to score recall and the criterion used to categorise prejudgement responses. A response was categorised as partially correct if it contained at least one of the idea units from the correct answer, whereas a recall score of half credit was assigned only if the response had 50% or more of the correct idea units. Thus, some responses were categorised as partially correct but received scores of 0 (e.g., when only one of four idea units was recalled). The logic of this categorisation scheme was to keep the commission category "pure", so that responses in this category contained no objectively correct information. Accordingly, responses that contained any correct information were included in the partial category, even if it did not contain enough to merit a half credit score.

standard group, recall significantly improved from prejudgement to criterion test for omissions and commissions, $ts > 2.24$, whereas recall significantly declined from prejudgement to criterion test for correct responses, $t(23) = 2.74$, $p < .05$. Recall for prejudgement and criterion test did not significantly differ for partially correct responses, $t(25) = 0.17$. Importantly, although performance gains in some categories were observed in both groups, greater gains were obtained in the standard group. In addition to the significant main effect of time of test, the interaction of time of test and group was significant for omissions, $F(1, 50) = 14.12$, $p < .001$, commissions, $F(1, 54) = 20.51$, $p < .001$, and partially correct responses, $F(1, 53) = 8.84$, $p < .01$; the interaction for correct responses was not significant, $F = 1.53$. Performance on the criterion test was greater for the standard group than for the no standard group for omissions, commissions, and partially correct responses, $ts > 3.41$.

These results show that providing an external standard not only improves the accuracy of self-score judgements, but it can also improve memory for the evaluated information. Moreover, just as the effects of providing a standard on self-score judgement accuracy depended on the kind of prejudgement recall response, the effects of providing a standard on memory also depended on the kind of prejudgement recall response. The effects of providing a standard on memory demonstrated here also contribute to the literature investigating the effects of retrieval practice and restudy on recall. We will consider these issues further in the Discussion.

Global and term-specific prediction accuracy

Although prediction accuracy is not relevant to empirically evaluating the focal hypotheses of this research, for interested readers and for archival purposes, we briefly report conventional analyses of prediction accuracy here. For each individual, we computed a gamma correlation between the six global predictions and criterion recall for each text. Means across individuals' correlations did not differ significantly between the group receiving the standard ($M = 0.36$, $SE = 0.09$) and the no standard group ($M = 0.21$, $SE = 0.13$), $t(52) = 0.99$.

For the term-specific predictions, we computed a gamma correlation across the 24 term-specific predictions and criterion recall for the corresponding terms for each individual. Means across individual correlations were reliably lower for the standard group ($M = 0.52$, $SE = 0.06$) than for the no standard group ($M = 0.68$, $SE = 0.05$), $t(53) = 2.01$. Although one may expect that predictive accuracy would improve with the presence of feedback, the feedback itself influenced criterion recall (as discussed above), which in turn would undermine predictive accuracy (see also, Kimball & Metcalfe,

2003). Perhaps most important, note that accuracy was moderate-to-high for term-specific predictions and higher than the accuracy of the global predictions, which replicates outcomes reported by Dunlosky et al. (2005).

DISCUSSION

The present study was motivated by previous research showing that when students are asked to assess their learning of key terms from expository text, the accuracy of these judgements is constrained. One way students could assess their learning is by attempting to recall the target definition and then by evaluating the *correctness* of the output of the retrieval attempt. The idea here is that this judgement involves a participant's evaluation of how well the retrieved information represents the actual meaning of the definition. In this way, these judgements tap metacomprehension because they involve evaluating one's understanding of a concept. Unfortunately, students' judgements about the correctness of retrieved definitions are only moderately accurate (Dunlosky et al., 2005), which suggests that judgement accuracy is partly constrained by inaccurate metacomprehension. In the present research, we tested two specific hypotheses for the limited accuracy of people's evaluations of the correctness of retrieved information—the limited competence hypothesis and the absence of standard hypothesis.

According to the limited competence hypothesis, students have limited ability to recognise inconsistencies between information they have recalled and objectively correct information, even when provided with an external standard for comparison. Results provided some support for this hypothesis. In particular, providing an external standard did not significantly improve the accuracy of self-score judgements for prejudgement recall responses that were partially or completely correct. Furthermore, although the provision of standards did improve judgements for commission errors somewhat, students who were provided an external standard still assigned partial or even full credit to 43% of responses that were completely incorrect.

Why might students be limited in the extent to which they can recognise inconsistencies between information they have recalled and objectively correct information? Two possibilities explored here were (a) that students' evaluations are unduly influenced by response quantity, and (b) that they did not consult the objectively correct information when evaluating their response. Secondary analyses suggested that the influence of both factors was minimal. Although not evaluated here, another plausible explanation for at least some of the inflated self-scores may be that students expect to receive some credit just for trying, based on prior classroom experiences involving similar evaluation situations. Although this explanation is unlikely to account for cases in which students awarded full credit to completely incorrect responses, it may account for some cases in which incorrect responses are assigned partial credit. Future research could evaluate this account by providing participants with explicit training and examples of how response correctness will be graded on the final test.

Another possibility is that the extent to which students can recognise inconsistencies between their responses and correct information is constrained by limited working-memory capacity. According to the construction-integration theory of text comprehension (Kintsch, 1988, 1998), text material is processed within a limited capacity processing system. Thus, text material is processed in cycles, with the amount of input processed in a cycle roughly equivalent to a simple sentence. Only a limited amount of information may be carried over from one processing cycle to the next, because some capacity must be available for processing the subsequent input to the processing system. Similarly, in order to recognise an inconsistency between a response and a standard, an individual must have both pieces of information concurrently active in working memory (Otero & Kintsch, 1992). Applied to the present task, students may not be able to carry over all of the content in their response to the next cycle in which the objectively correct information is processed (or vice versa). Although external presentations of one's response and the correct standard may reduce the capacity demands of making a comparison, visual comparison of the two pieces of information is not enough. That is, even though they may contain some of the same surface information (e.g., words or syntax), such overlap does not ensure that they will contain the same semantic information. Thus, the semantic content of the response and the correct standard must be processed together, which may exceed the capacity of the working memory system in which the processing of semantic information takes place.

Importantly, in addition to the evidence for the limited competence hypothesis, we also found some support for the absence of standard hypothesis. According to this hypothesis, evaluation of the correctness of recall output will be constrained when students do not have an external standard to which their output can be compared. Consistent with this hypothesis, providing an external standard improved the accuracy of self-score judgements. Although students in the standard group were still inappropriately assigning some credit to incorrect responses 43% of the time, this bias was significantly less than the 83% error rate observed in the no standard group (Table 1).

Effects of standards on memory

In addition to improving the accuracy of self-score judgements, providing an external standard also improved memory for the target information.

The finding that criterion recall was greater in the standard group than in the no standard group is consistent with previous research on the effects of retrieval practice and restudy. For example, Cull (2000) presented individuals with vocabulary word pairs (e.g., "handsel–payment") for an initial study trial. The word pairs were then presented three more times in one of three conditions: Individuals either attempted to retrieve the second word of a pair when presented with the first word as a cue, restudied the word pair, or both (i.e., a retrieval attempt followed by presentation of the correct answer for restudy). Performance on a final cued recall test was greater after retrieval plus restudy than after retrieval alone or restudy alone.

Consistent with the current results, recent research has further suggested that the efficacy of restudy after retrieval for improving memory depends on the nature of the initial retrieval response. Pashler, Cepeda, Wixted, and Rohrer (2005) presented individuals with Luganda–English translation equivalents (e.g., "leero–today") for two study trials. Cued recall for the vocabulary pairs was then tested twice. One group did not receive feedback about the accuracy of their responses, whereas a second group was provided with the correct answer after each retrieval attempt. Recall on the second test was greater when correct answers had been provided during the first test than when no answers were provided. However, this effect was only significant for responses that were omissions or commissions on the first test; the two groups did not significantly differ in criterion recall for initially correct responses. These results mirror the present findings: Criterion recall was greater in the standard group than in the no standard group, but only for those items that were not correctly recalled during prejudgement recall. In this way, the present research provides an important advance beyond previous research on retrieval practice and restudy, by showing that these effects generalise from simple verbal materials (word pairs) to more complex text materials. Our results converge with those reported by Kang, McDermott, and Roediger (2007 this issue), who also found that feedback provided after short answer practice tests was particularly beneficial for initially incorrect responses. However, we should note that Kang et al. did not find this effect of feedback following multiple choice practice tests. Furthermore, Butler and Roediger (2007 this issue) did not find any effect of feedback for either kind of practice test, although this may have been due to a relatively high rate of initially correct responses. In short, further exploring the effects of feedback on memory for more complex materials is an important direction for future research.

Educational implications

What are the potential implications of these results for promoting the effectiveness of students' self-regulated learning? Given that self-testing is often implicitly or explicitly recommended to students, the present research suggests that self-testing may be an effective strategy for improving students' self-regulated learning when a standard for evaluating the outcome of the self-test is available. First, providing a standard that consists of the correct answer can improve memory for the to-be-learned content. Second, providing a standard also improves the accuracy of students' evaluations of their own learning, which may further improve learning by supporting effective self-regulation of subsequent study. To effectively regulate study, students must be able to accurately identify material that they have not learned well enough so that subsequent study can be focused on that content. Of course, we are not suggesting that accurate monitoring is sufficient for effective self-regulation, because students must also be motivated to study and appropriately use the output from monitoring to guide study. Nevertheless, the finding that providing a standard improves the accuracy of self-score judgements suggests that it could enhance the effectiveness of self-regulated study. Consider again the results presented in Table 1. Assuming that students would choose to restudy any item for which their self-test response was partially or completely incorrect, students who were provided with a standard would only have failed to select 6% of their incorrect responses for restudy. Without a standard, 25% of the incorrect responses would not have been studied further.

The present results also have implications for instruction. Textbooks often contain end-of-chapter review questions (e.g., a list of the key concepts covered in that chapter), and instructors often provide students with study guides to prepare for exams. The implicit or explicit instruction to students is that they should use these materials as a basis for self-testing. Our previous research suggests that students may not always self-test their memory even when provided with cues that afford retrieval practice. Instead of evaluating memory using a full-blown retrieval attempt, they may assume they know the answer if they feel familiar with the terms themselves (e.g., Reder & Ritter, 1992). The present research further suggests that even when students do self-test, the effectiveness of the strategy for evaluating and improving learning will depend critically on whether an external standard is provided for comparison to the outcome of the self-test. Unfortunately, although many textbooks provide key term lists at the end of chapters, they often do not provide the correct responses in a concise and easy-to-reference manner. Of course, students could look back through the chapter to find each definition to check against their self-test response, but this would obviously be cumbersome and time consuming. Likewise, study guides provided by instructors are rarely accompanied by corresponding answers. Instructors may assume that students will revisit lecture notes to find the correct information for comparison to self-test responses. However, students' lecture notes are often grossly incomplete and lacking in organisation (e.g., Titsworth & Kiewra, 2004), and thus it will often be difficult or impossible for students to locate standards on their own. Accordingly, one way in which instructors can support more effective student self-regulated learning is to provide external standards for use during self-testing.

Although we suggest that self-testing followed by an external standard for evaluating the outcome can be an effective strategy for improving students' self-regulated learning, we do not mean to imply that the self-testing strategy is sufficient for meeting learning goals in all situations. Indeed, for much of the content in many courses, the expectation is that students will be able to think critically about and apply the information they learn. Accurate evaluation of one's self-testing outcomes against an external standard can improve memory for that information and inform subsequent restudy decisions to further improve learning, but it does not guarantee that students will be able to think critically about or apply what they are recalling. Nevertheless, the importance of being able to remember key terms and concepts should not be diminished: Students will not be able to apply a concept they cannot remember, nor think critically about those they cannot recall. Thus, the self-testing strategy may benefit students' comprehension of key concepts at least indirectly and allow them to more efficiently meet their learning goals.

REFERENCES

Baker, J., & Dunlosky, J. (2006). Does momentary accessibility influence metacomprehension judgments? The influence of study-judgment lags on accessibility effects. *Psychonomic Bulletin and Review, 13*, 60–65.

Benjamin, A. S., Bjork, R. A., & Schwartz, B. L. (1998). The mismeasure of memory: When retrieval fluency is misleading as a metamnemonic index. *Journal of Experimental Psychology: General, 127*, 55–68.

Butler, A. C., & Roediger, H. L., III. (2007). Testing improves long-term retention in a simulated classroom setting. *European Journal of Cognitive Psychology, 19*, 514–527.

Cull, W. L. (2000). Untangling the benefits of multiple study opportunities and repeated testing for cued recall. *Applied Cognitive Psychology, 14*, 215–235.

Dunlosky, J., & Rawson, K. A. (2005). Why does rereading improve metacomprehension accuracy? Evaluating the levels-of-disruption hypothesis for the rereading effect. *Discourse Processes, 40*, 37–55.

Dunlosky, J., Rawson, K. A., & McDonald, S. L. (2002). Influence of practice tests on the accuracy of predicting memory performance for paired associates, sentences, and text material. In T. Perfect & B. Schwartz (Eds.), *Applied metacognition* (pp. 68–92). Cambridge, UK: Cambridge University Press.

Dunlosky, J., Rawson, K. A., & Middleton, E. L. (2005). What constrains the accuracy of metacomprehension judgments? Testing the transfer-appropriate-monitoring and accessibility hypotheses. *Journal of Memory and Language, 52*, 551–565.

Hacker, D. (1998). Self-regulated comprehension during normal reading. In D. J. Hacker, J. Dunlosky, & A. C. Graesser (Eds.), *Metacognition in educational theory and practice* (pp. 165–191). Hillsdale, NJ: Lawrence Erlbaum Associates, Inc.

Howe, M. J. A. (1970). Repeated presentation and recall of meaningful prose. *Journal of Educational Psychology, 61*, 214–219.

Johnson, H. M., & Seifert, C. M. (1994). Sources of the continued influence effect: When misinformation in memory affects later inferences. *Journal of Experimental Psychology: Learning, Memory, and Cognition, 20*, 1420–1436.

Kang, S. H. K., McDermott, K. B., & Roediger, H. L., III. (2007). Test format and corrective feedback modify the effect of testing on long-term retention. *European Journal of Cognitive Psychology, 19*, 528–558.

Kelemen, W. L. (2000). Metamemory cues and monitoring accuracy: Judging what you know and what you will know. *Journal of Educational Psychology, 92*, 800–810.

Kimball, D. R., & Metcalfe, J. (2003). Delaying judgments of learning affects memory, not metamemory. *Memory and Cognition, 31*, 918–929.

Kintsch, W. (1988). The role of knowledge in discourse comprehension: A construction-integration model. *Psychological Review, 95*, 163–182.

Kintsch, W. (1998). *Comprehension: A paradigm for cognition.* Cambridge, UK: Cambridge University Press.

Koriat, A. (1993). How do we know that we know? The accessibility model of the feeling of knowing. *Psychological Review, 100*, 609–639.

Maki, R. H. (1998). Test predictions over text material. In D. J. Hacker, J. Dunlosky, & A. C. Graesser (Eds.), *Metacognition in educational theory and practice* (pp. 117–145). Hillsdale, NJ: Lawrence Erlbaum Associates, Inc.

Miesner, M. T., & Maki, R. H. (2007). The role of test anxiety in absolute and relative metacomprehension accuracy. *European Journal of Cognitive Psychology, 19*, 650–670.

Morris, C. C. (1990). Retrieval processes underlying confidence in comprehension judgments. *Journal of Experimental Psychology: Learning, Memory, and Cognition, 16*, 223–232.

Nelson, T. O., Narens, L., & Dunlosky, J. (2004). A revised method for research on metamemory: Pre-judgment recall and monitoring (PRAM). *Psychological Method, 9*, 53–69.

Otero, J., & Kintsch, W. (1992). Failures to detect contradictions in a text: What readers believe versus what they read. *Psychological Science, 3*, 229–235.

Pashler, H., Cepeda, N. J., Wixted, J. T., & Rohrer, D. (2005). When does feedback facilitate learning of words? *Journal of Experimental Psychology: Learning, Memory, and Cognition, 31*, 3–8.

Reder, L. M., & Ritter, F. E. (1992). What determines initial feeling of knowing? Familiarity with question terms, not with the answer. *Journal of Experimental Psychology: Learning, Memory, and Cognition, 18*, 435–451.

Son, L. K., & Metcalfe, J. (2005). Judgments of learning: Evidence for a two-stage process. *Memory and Cognition, 33*, 1116–1129.

Thiede, K. W. (1999). The importance of monitoring and self-regulation during multi-trial learning. *Psychonomic Bulletin and Review, 6*, 662–667.

Thiede, K. W., Anderson, M. C. M., & Therriault, D. (2003). Accuracy of metacognitive monitoring affects learning of text. *Journal of Educational Psychology, 95*, 66–73.

Titsworth, B. S., & Kiewra, K. A. (2004). Spoken organizational lecture cues and student notetaking as facilitators of student learning. *Contemporary Educational Psychology, 29*, 447–461.

Weaver, C. A., Bryant, D. S., & Burns, K. D. (1995). Comprehension monitoring: Extensions of the Kinstch and van Dijk model. In C. A. Weaver, S. Mannes, & C. R. Fletcher (Eds.), *Discourse comprehension: Essays in honor of Walter Kintsch* (pp. 177–193). Hillsdale, NJ: Lawrence Erlbaum Associates, Inc.

EUROPEAN JOURNAL OF COGNITIVE PSYCHOLOGY
2007, 19 (4/5), 580–606

Retrieval-induced forgetting in educational contexts: Monitoring, expertise, text integration, and test format

Marie Carroll, Jonathan Campbell-Ratcliffe, and Hannah Murnane

Australian National University, Canberra, Australia

Timothy Perfect

University of Plymouth, Plymouth, UK

Retrieval practice leads to the impaired recall of related but unpractised items, an effect termed retrieval-induced forgetting (RIF). Two experiments showed that RIF occurred with "real world" educational material, and isolated several boundary conditions for the phenomenon. Experiment 1 showed that integration of material available to experts but not to novices in a domain of knowledge, protected against RIF, which disappeared after a 24 hours. Experiment 2 examined the impact on RIF of the degree of coherence or integration of the text material itself and the type of test format administered. Text coherence did not influence RIF, which occurred for the short answer and essay tests, but not for the multiple choice test. In both experiments, those participants who demonstrated RIF were able to monitor accurately the likelihood of recall impairment, suggesting that RIF may not be an unconscious process. Results are discussed in relation to exam preparation strategies.

When competition is introduced between items that are to be learned, participants can be unconsciously induced to forget those items that compete for retrieval (Anderson, Bjork, & Bjork, 1994). For example, if participants are given two categories of items to learn, and then perform sustained retrieval practice on half of the items from one of the categories, their later recall of unpractised competitors from the practised category is found to be impaired, relative to their recall of items from the wholly unpractised category. This form of unconscious forgetting is termed RIF, and has been found to be quite robust across a large range of studies.

Correspondence should be addressed to Marie Carroll, QESS, Australian National University, Canberra, ACT 0200, Australia. E-mail: marie.carroll@anu.edu.au

http://www.psypress.com/ecp   DOI: 10.1080/09541440701326071

Investigations of this effect have touched on several important practical implications, ranging from how students study for exams (and may unwittingly forget studied information), to the effect of interviews (and thus retrieval practice) on eyewitness testimony, to the effects of forgetting in social psychology. For instance, Macrae and MacLeod (1999) demonstrated that RIF could be elicited under mock examination conditions, while MacLeod (2002) and Shaw, Bjork, and Handal (1995) found RIF was the cause of impaired recall in a modified eyewitness paradigm.

Many studies (including Anderson, Bjork, & Bjork, 2000; Anderson & Spellman, 1995; Bäuml & Kuhbandner, 2003; Dunn & Spellman, 2003; Johnson & Anderson, 2004; Perfect, Moulin, Conway,& Perry, 2002; Saunders & MacLeod, 2002; Shivde & Anderson, 2001; and see Levy & Anderson, 2002, for a review) have argued that RIF can be best accounted for by inhibitory processes. Specifically, it is argued that while repeated retrieval practice clearly leads to the strengthening of practised items, those items that are semantically related, and will later compete for activation during retrieval, are inhibited to better improve the recall of those items that are most appropriate for completing the task.

The cue used to elicit inhibited items may be varied without affecting the inhibition effect. As shown by Anderson and Spellman (1995), the cue–target relationship alone is not the issue—forgetting is highly dependent on the competing items, and specifically their contextual relationship to the practised items.

The present studies are concerned to investigate the occurrence of RIF in formal classroom learning, especially in relation to students studying for exams. There are several important boundary conditions for RIF that have particular implications in classroom learning. The first boundary condition is the influence of integration of learned material on RIF, the second is the duration of the RIF effect, and the third is the effect of type of test format that is most susceptible to RIF. Together, these boundary conditions may make RIF more or less likely to be observed in classroom learning, and are the focus of the investigation in the following studies.

INTEGRATION OF LEARNED MATERIAL

A factor that appears to moderate the amount of RIF experienced is the degree to which participants are able to integrate the practised items into a cohesive whole. Generally speaking, integration refers to the existence of interconnections between items sharing a common retrieval cue. These interconnections can be long standing (i.e., existing preexperimentally), or novel (i.e., generated during the course of an experiment). Studies of fact recognition initially suggested that integration may protect against RIF. For example, Radvansky and Zacks (1991) instructed participants to learn a series of facts (e.g., "the palm is in the lobby", "the ashtray is in the lobby"). Generally, the more facts the participants learned, the slower their recall of specific facts (termed the "fan effect"). But if participants are able to integrate these facts (in the above example, the facts are grouped by their common location), little interference is found. Several studies indicate that integration can also reduce RIF, and, interestingly, it appears that little needs to be done to encourage participants to integrate items while studying and performing retrieval practice. Anderson and McCulloch (1999) found, in a simple modification of the Anderson et al. (1994) paradigm, that an initial instruction to "integrate information" was enough to significantly reduce the amount of RIF observed during the final test phase. When Anderson and McCulloch later divided their participants into high and low integra- tors, depending upon their responses in postexperimental questionnaires, they found that participants who reported that they had spontaneously attempted to integrate the material, even when not specifically required to, also demonstrated significantly less RIF. In a further demonstration, Anderson and Bell (2001) found that integration through the formation of single vivid mental images protected against the forgetting of propositional materials (e.g., "the ant crawled on the table", "the ant crawled on the rock"). However, it is important to note that while both of these studies found that participants who integrated the material demonstrated a reduction in the amount of RIF, it rarely completely eliminated it. Rather, the amount of integration reported by participants tended to be associated with a proportionate reduction in the amount of RIF observed during the final testing phase.

The findings of Dunn and Spellman (2003) also favour the role of integration as a boundary condition for RIF. In this study, while practising some stereotypic traits (e.g., Artist–creative, or Asian-American–intelligent) inhibited the recall of unpractised traits relating to that stereotype, this effect was strongly modified by the degree to which the participant believed in the stereotype. Dunn and Spellman argued that, as a stereotype provides an integrative framework for understanding the world, participants who strongly believed in the validity of the stereotype were able to cohesively integrate the stereotyped traits. This integration reduced the amount of interference between competing traits, thus reducing the degree of inhibition and the amount of forgetting.

As Anderson and McCulloch (1999) point out, the effect of RIF is somewhat paradoxical when we consider expertise. Assuming that learning more information about a topic increases competition, then as we acquire more expertise within a particular area we should become *less* adept at retrieving items from within this domain. If the inhibitory processes behind RIF represent an adaptive attempt to reduce competition, then surely an expert's large domain of knowledge should be particularly prone to inhibition and thus forgetting. However in reality, it seems that experts generally have *better* access to information from within their respective domains. In an attempt to understand this paradox, Smith, Adams, and Schorr (1978) proposed that experts may integrate knowledge from within their domain, and this integration may be what protects against forgetting. Experts do not accumulate new facts in isolation, but rather integrate them within a rich body of background knowledge that supports an understanding of the many interrelations between items. Anderson and McCulloch (1999) argue that this integrative framework encourages the understanding of interrelations between items and the development of novel retrieval strategies, which may protect experts from the paradoxical nature of RIF. To date, however, there have been few studies of the operation of RIF using complex information in experts' domains of knowledge. One such study recently conducted (Barnier, Hung, & Conway, 2004) did find RIF for emotional and unemotional autobiographic memories, knowledge, which could be construed as coherent and organised in the same way that expertise is. However, further investigation is necessary to determine whether high levels of integration in experts are capable of reducing competition between items, and thus protect them from RIF in classroom settings.

One study that has examined the educational implications of RIF is that of Macrae and MacLeod (1999), who subjected participants to a mock exam scenario, in which they were presented with facts relating to two fictional tropical islands (Tok and Bilu). They then performed retrieval practice on half of the facts relating to one of the islands. Consistent with RIF, recall of unpractised facts about the practised island was impaired relative to the recall of facts relating to the unpractised baseline island. However, in this study the "fact" to be learned was explicit, with equal weight given to each (i.e., "Bilu's only major export is copper"). There was also no rich fabric of background knowledge on which to draw to support learning. In contrast, in real-world learning contexts, it is up to the student to read through a large body of information, to draw inferences based on background and textual knowledge, and to select ideas likely to be critical (and likely to be tested later) from those which are less important. It is unclear whether RIF would occur under these circumstances.

DURATION OF THE RIF EFFECT

The literature consistently agrees that inhibition in the retrieval-practice paradigm will be evident from 5 (Macrae & MacLeod, 1999) to 20 min (Anderson et al., 1994, among others) after retrieval practice is performed.

Several studies have indicated that RIF will recover over time. MacLeod and Macrae (2001, Study 1) tested participants either 5 min or 24 hours after performing retrieval practice, and while RIF was observed immediately following retrieval practice, it was not evident after 24 hours. This result indicates that somewhere in this time period, the inhibition of competing items was reduced to zero. In their Study 2, the authors placed a 24 hour delay between either retrieval practice and the final test, or between study and retrieval practice. Interestingly, this manipulation produced the expected forgetting only among participants in the second condition, indicating that a delay of 24 hours only eliminated RIF when it occurred *after* retrieval practice, and not after encoding, although the magnitude of forgetting was not as great as when all phases followed closely.

METAMEMORY MONITORING

A further unexplored issue is whether individuals are aware of the memory impairments that RIF generates. Dunn and Spellman (2003) argue that lateral inhibition is responsible for the RIF effect, and is automatically activated when attention is drawn to one category or another. Unlike hierarchical inhibition, which involves the conscious and intentional inhibition of items via top-down methods of processing (thought to be responsible for directed forgetting), it is an automatic process. Even so, this does not necessarily imply that it is always outside consciousness. Surprisingly, to our knowledge, no research has investigated self-monitoring of RIF—specifically whether individuals are aware that performing retrieval practice on some items may impair the later recall of related but unpractised items. There are educational implications here. If students are aware that once-learned but unpractised information is likely to be inhibited, is there anything they can do to compensate for this? The very retrieval practice that students are repeatedly encouraged to perform before tests—and which has been shown to improve test performance (e.g., Roediger & Karpicke, 2005)—may make their memory for some items worse. It is normally impracticable to rehearse all of the information one needs to know about a topic in preparing for an exam; usually only a subset is subject to retrieval practice. Yet there may be an unwarranted assumption on the learner's part that recently learned unrehearsed information related to recently retrieved information is as likely as any other once-learned information to be remembered. Although performing compensatory activities may not be an option, it is at least useful to educators to know whether learners are aware of the inhibitory effects on unpractised items. Research has demonstrated that individuals revise learning strategies in light of the content they feel has and has not been consolidated into memory (Nelson & Leonesio, 1988;

Pelegrina, Bajo, & Justicia, 2000); yet, it is unclear whether memory inhibition may influence these judgements. Judgements of the success and progress of an individual's own learning are functions of an individual's metacognitive monitoring (Nelson & Narens, 1994; Schwartz & Perfect, 2002). One such judgement is the Judgement of Learning (JOL), typically an assessment made after the learning phase and prior to final testing, which aims to assess people's degree of confidence that they will correctly recall the information that they have learned at a later time (Nelson & Narens, 1994). If individuals in the retrieval-practice paradigm indicate that they are less confident in their ability to recall the unpractised items (Rp−) from the practised category (Rp+) than to recall unpractised items from the baseline category (NRp items), this may indicate that RIF is not a wholly unconscious phenomenon.

Our primary aim in these studies was to investigate whether integration can protect against forgetting. In the first study we examined the hypothesis put forward by Anderson and McCulloch (1999) that experts may be protected against forgetting information from their domain of expertise through an ability to integrate new information into a cohesive framework. In the second study we investigate integration by varying the type of material presented. In both cases, the material used was real-world classroom text to be learned for an upcoming test.

EXPERIMENT 1

In Experiment 1, the "experts" were students who had studied psychology formally at university for 4 or more years; the novices were those in the first year of university psychology study. The two case studies were taken from university textbooks describing schizophrenia and autism respectively (but with no reference to these labels). It was thought that experts would readily integrate the symptom set into one of the two disorders, while novices would be less able to label and, therefore, integrate the symptom set into a preexisting schema for the disorder, though they may recognise a subset of symptoms as typical of one or the other disorder.

We examined the occurrence of RIF for such material when testing took place immediately and when it was delayed by 24 hours. While MacLeod and Macrae (2001) found that RIF did not occur when retrieval practice and final testing were separated by 24 hours, the duration of inhibition may depend greatly upon the type of material that is being inhibited (Anderson, 2003). Category-exemplar pairs may be inhibited for a different length of time than real-world material, which may in turn be different from that found for the simple, artificial "facts" of the MacLeod and Macrae study. To date no study has examined the effect of RIF on complex real-world material, and thus its persistence is unknown. For practical purposes, a finding that persistence of forgetting occurs after 24 hours is of concern to students and teachers, since many students study only in the 24-hour period before an exam.

Additionally, Experiment 1 included JOL ratings for the material that had been learned. These were elicited following the retrieval-practice phase for both rehearsed and unrehearsed items in order to determine whether learners were able to monitor RIF, should it occur.

Method

Participants and design

There were 32 participants in each of two categories: novices and experts. The novices, half of whom were first year Psychology students, had a mean of 0.38 years of psychology experience. For the expert participants the mean number of years of psychology experience was 4.55 ($SD = 0.76$), and all were Psychology majors.

This experiment involved a $2 \times 2 \times 2$ mixed factors design, where type of item (two levels, Rp− vs. NRp[1]) was manipulated within subjects, while the degree of expertise (two levels: novice vs. expert) and the time period between retrieval practice and final testing (two levels: immediate vs. 24 hours) were between-subject factors. The dependent variables were: (1) the number of questions answered correctly and (2) the mean JOL rating.

Materials

Two case studies with 40 related questions were constructed, with content drawn largely from a third-year abnormal psychology textbook. One case study was of a schizophrenia sufferer named David, and the other was of an autism sufferer named Peter. The texts were rewritten as vignettes apparently describing particular cases, so as to ensure that all the information was entirely novel, and could not be predicted without having first read the papers. For each case study, 20 specific and distinct questions were devised (see Appendix A for examples). An effort was made to include a range of easy and difficult questions, but the emphasis was on including more difficult questions to eliminate ceiling effects amongst experts. During the final testing phase, each participant was presented with only 30 of the total 40 questions. These 30 comprised 10 practised questions (Rp+), 10 unpractised questions from the practised case study (Rp−), and 10 baseline questions from the unpractised case study (NRp).

---

[1] An explanation of NRp and Rp− follows in the Procedure section.

Participants were provided with an answer booklet, with two variations. The first pages consisted of 30 rating scales intended for the assessment of JOLs. These scales ranged from 0 to 100 in increments of 20, with 0 indicating no confidence of recalling the answer at final testing, and 100 indicating full confidence. In the immediate condition a distractor task followed, comprising three pages containing each letter of the alphabet written sequentially, and a category heading at the top of the page. This section was absent from answer booklets used in the 24-hour condition. On the following pages, numbered spaces allowed participants to write their answers for all 30 questions in the final testing phase.

Procedure

Participants were randomly assigned to one of the testing conditions. The following procedure follows the original Anderson et al. (1994) retrieval-practice paradigm closely, but with the addition of a JOL phase.

Initial study phase.   Participants were instructed to read the two study papers carefully. It was made clear that they would be tested on their comprehension of these papers later with a series of short answer questions, and were instructed to "pick out any points which they considered salient". Presentation of the papers was counterbalanced—half of the participants read the David/schizophrenia paper first, while half read the Peter/autism paper first.

Retrieval-practice phase.   Participants were then read a series of 10 questions, which they were told would appear on the final test. This phase was intended to mimic the practice, which is common prior to an exam, where examinees quiz one another on material that they may be tested on. Participants were instructed to respond verbally with the answer if they knew it, and to guess even if they were unsure. If they were incorrect or did not know the answer, they were required to repeat the correct answer, which was provided by the experimenter. This repetition was critical as it was intended to ensure retrieval of the material, as opposed to simple re-presentation, which may not have produced the desired RIF effect (Bäuml, 2002). Participants practised half of the questions from their assigned set, and each question was repeated three times. Participants were randomly assigned to practise either the odd or even questions from one of the sets for counterbalancing purposes. These practised questions became the Rp+ items for that participant, the unpractised questions from the same case study became their Rp− items, and 10 odd or even items from the unpractised case study became their NRp items.

Judgements of learning. Next, participants were shown each question, without the answer, that would appear in their final test, and for each they were instructed to indicate their level of confidence that they would correctly recall the answer during the final testing phase. It was emphasised to participants that they should avoid thinking of the answers themselves, to avoid a retrieval practise opportunity for the unpractised items. To reduce the likelihood of covert retrieval, participants were instructed to respond as quickly as they could, and were reminded of this every time their response was delayed.

Distractor phase. Those participants assigned to the "immediate" condition engaged in a series of distractor tasks, lasting 15 min, which involved generating 26 words beginning with each letter of the alphabet, for three categories. Participants in the "24-hour" condition were instructed to return at the same time on the following day for the final phase of the experiment.

Final test phase. Participants were provided with a question booklet containing their 30 counterbalanced questions, and were instructed to answer as many as they could in their answer booklet. Participants were told that they would not lose any marks for incorrect answers, and thus to try to answer as many questions as they could.

Results

Initial screening. A significance level of .05 was used for all statistical tests. It was possible that certain combinations of questions were easier or harder than others, which could affect the amount of inhibition demonstrated by participants with these combinations of questions. To determine if this was the case, two one-way ANOVAs were conducted, the first assessing whether the choice of practised case study affected the amount of inhibition, and the second assessing whether practising odd or even questions affected the amount of inhibition. Neither was found to be significant, $F(1, 62) = 3.081$, $p = .084$; and $F(1, 62) = 0.391$, $p = .534$, respectively.

Proportion correct. Retrieval practice did facilitate later recall, as can be seen from the means for Rp+ items in Table 1. Because the large, predicted practice effect swamps other effects, the analyses that follow did not include Rp+ items.

Several researchers have argued that RIF is measured as the amount of inhibition each participant demonstrates (see, for example, Perfect et al., 2002). Inhibition is calculated for each participant by subtracting Rp− scores from NRp scores separately for each condition. Table 1 shows the

TABLE 1
Mean proportion of questions answered correctly in each experimental condition in Experiment 1

Condition	Rp +	Rp −	NRp	Inhibition (RP −)−(NRp)
Novice/immediate	.894	.338	.506	−.169
Novice/24 hour	.763	.356	.394	−.038
Expert/immediate	.956	.525	.575	−.050
Expert/24 hour	.869	.519	.538	−.019

mean scores on amount of inhibition for each of the four conditions. The 2 (novice vs. expert) × 2 (immediate vs. 24 hour) × 2 (Rp − vs. NRp) ANOVA showed that, not surprisingly, experts performed better (mean = 0.53, standard error = 0.28) than novices (mean = 0.39, standard error = 0.29), $F(1, 60) = 11.98$, $p < .001$, $MSE = 5.28$, and that NRp items (mean = 0.50, standard error = 0.22) were answered more correctly than Rp − (mean = 0.43, standard error = 0.22), $F(1, 60) = 13.18$, $p < .001$, $MSe = 15.12$, confirming an overall RIF effect. The main effect of delay did not reach significance (mean immediate = 0.48; mean 24 hour = 0.45). The interaction between the two factors—novice/expert and type of question—did not reach significance, $F = 3.29$, $p > .05$; however, the levels of inhibition presented in Table 1 suggest that novices experienced greater RIF than experts in the immediate condition. Experts, then, are protected somewhat from RIF.

There was also a significant interaction between delay and type of question, $F(1, 60) = 4.61$, $p < .05$, $MSe = 1.47$; after 24 hours the RIF effect was much less pronounced than in the immediate condition. The three-way interaction was not significant.A separate analysis was carried out using inhibition (the mean difference, for each subject, between Rp − and NRp) as the dependent variable. As it was hypothesised that inhibition would be greater in the novice/immediate condition, a simple planned comparison was executed. The resulting $F(3, 60) = 9.303$, $p < .05$ and $R^2 = .134$ indicated that the amount of inhibition in the novice/immediate condition did indeed differ significantly from the other three conditions, supporting the hypothesis that RIF could be produced for this naturalistic material.

As it was specifically predicted that novice participants in the immediate condition would demonstrate greater inhibition than novices in the 24-hour condition, a final planned comparison was executed. This revealed that there was indeed a significant difference between these conditions, with $F(3, 60) = 6.007$, $p < .05$ and $R^2 = .091$, indicating significantly more inhibition amongst participants in the immediate condition.

TABLE 2
Mean proportion of JOL ratings by item type for each condition, Experiment 1

Condition	JOL +	JOL −	JOLN	Inhibition (JOL −)−(JOLN)
Novice/immediate	.94	.52	.56	−3.50
Novice/24 hour	.79	.53	.60	−6.25
Expert/immediate	.98	.70	.65	+4.50
Expert/24 hour	.98	.68	.69	−1.25

Magnitude of judgements of learning. A similar 2 (novice/expert) × 2 (delay) × 2 (Rp −/NRp) ANOVA was carried out on the judgements of learning; the means are shown in Table 2. Again, items that had been practised (Rp +) received much greater JOL ratings; for this reason Rp + items were not included in the analysis. There were no significant main effects or interactions for JOL ratings. The Novice/expert × Type of item interaction failed to reach significance, $F = 3.53$, $p = .06$.

Normally, gamma correlations between the JOL rating and the response accuracy are calculated for each subject. However, with only 10 questions per condition (Rp +, Rp −, and NRp) there were many empty cells, and the gamma correlations were not able to be calculated.

It appears, from the JOL magnitude results that participants were not able to predict the significant inhibition that occurred; that is, they did not know that Rp − items would be answered less accurately than NRp items. Nevertheless, though the interaction was not significant, the means suggest that the novices were more likely to predict the inhibition effect than the experts. Perhaps, then, those subjects in general who are more susceptible to RIF are more likely to predict a poorer performance on Rp − than on NRp items.

To assess this possibility, participants were divided into two groups—those who demonstrated inhibition (those whose Rp − scores were lower than their NRp scores), and those who did not. Of the total 64 participants, 36 (or 56%) demonstrated some degree of inhibition. The mean proportion for JOLs for each question type across these two groups is presented in Table 3.

The amount of predicted inhibition (lower ratings to Rp − than to NRp items) was compared to determine whether those participants who demonstrated RIF also predicted this inhibition as shown by the proportions of JOLs. This was indeed found to be the case, $F(1, 62) = 11.22$, $p < .001$, $Mse = 85.15$, indicating that significantly more inhibition was predicted by those participants who demonstrated retrieval-induced forgetting.

TABLE 3
Mean proportion of JOLs for each question type for those participants who did and did not demonstrate RIF, Experiment 1

	n	JOL+	JOL−	JOLN	Inhibition (JOL−)−(JOLN)
Ps with inhibition	36	.94	.57	.63	−6.4
Ps without inhibition	28	.91	.66	.62	+4.57

Discussion

This study aimed to improve our overall understanding of the mechanisms involved in RIF, and to provide fresh directions for the development of learning strategies in real educational settings based upon this information. Our first goal was to determine whether RIF would occur for "real-world" material in a mock exam scenario, as previous research has focused on forgetting in laboratory settings using fairly artificial materials. Our second was to assess the notion of Anderson and McCulloch (1999): that integration may afford experts some protection from the detrimental effects of inhibition. Our third was to determine whether inhibition of this material would decay after 24 hours, as previous research has demonstrated (MacLeod & Macrae, 2001; Saunders & MacLeod, 2002). And finally, we sought to determine whether individuals have some level of awareness of the inhibition incurred by retrieval practice in this paradigm, or whether forgetting is entirely unconscious.

Each of these hypotheses was supported by the results. As was expected, the only condition that demonstrated significant levels of forgetting was the novice/immediate condition—participants in the other conditions were protected from forgetting by the effects of integration, or a sufficient time interval, which allowed the inhibition to decay. These results strongly suggest that RIF will occur for this real world material, but with two notable exceptions: (1) Domain knowledge affords protection from forgetting when the material can be integrated into a preexisting framework, and (2) a gap of 24 hours between retrieval practice and final testing is sufficient to reduce the forgetting of Rp− items to negligible levels. More intriguingly, these results also indicate that those individuals who experience inhibition may have some awareness of the detrimental effects of retrieval practice. This implies that the inhibition produced by retrieval practice may not be entirely unconscious. We return to this point in Experiment 2.

When viewed as a whole, the significantly smaller amounts of inhibition observed in the expert/immediate condition compared to the novice/immediate condition implies an advantage for those with some background in the material. There is some uncertainty about whether this advantage is directly due to integration of the material. Anderson and McCulloch (1999) observe that experts are better able to recall individual items from within a comparatively larger body of information, and while they argue that integration is the mechanism responsible, "integration" is a construct that is difficult to tap into directly, leaving the possibility that other models may also account for the reduced levels of forgetting observed among experts. For instance, it could be argued that experts have simply had more practice at recalling items from within their domain, and thus their improved recall of individual items may be a result of more frequent practice at recalling this type of material. Indeed, the experts have simply had more practice in study and test-taking in any domain; it is possible that it is the experience at university, rather than the domain expertise, that confers the protection. Future studies might aim to exercise more control over "expertise" using non-Psychology, senior students as experts.

Previous studies (MacLeod & Macrae, 2001; Saunders & MacLeod, 2002) have found that RIF is almost eliminated after 24 hours. This study provided support for these findings, with no significant amounts of inhibition evident after a delay of 24 hours between retrieval practice and final testing. However, it is possible that different types of material may produce different levels of inhibition that may persist for differing lengths of time. Presenting material in prose format may encourage integration over material presented as disconnected sentences; it is possible that the rate at which inhibition recovers may not be the same for these different types of material.

Interestingly, while our results have indicated that RIF is eliminated by a 24-hour break between retrieval practice and testing, facilitation of practised items when compared to nonpractised items was still apparent after this period. This observation raises an interesting question: Why is it necessary to inhibit competing items to facilitate the recall of practised items in the short term, yet after a 24-hour delay the facilitation of practised items remains but without the inhibition of unpractised items? It is possible that different mechanisms are involved in these differing time periods. Thus inhibition may represent a relatively short-term strategy for improving the recall of items against their stronger competitors, but when given a significant temporal delay, the relative strengthening of the practised items alone is sufficient to improve their recall whilst avoiding the false recall of competing items.

It is clear that material of this type is not immune to the detrimental effects of retrieval practice. While RIF is an issue neither for long-term learning nor for material being learned within a domain of expertise, it is still the case that in some classroom learning situations, new material may be learned, practised, and finally quizzed within a single learning session. Such a teaching practice would be highly inadvisable based upon these findings, as it would almost certainly result in significant levels of RIF.

EXPERIMENT 2

Experiment 1 found that RIF is a potentially serious limitation on test performance, at least for novices in a knowledge domain, and that its effect is most serious immediately after learning. Nevertheless, learners do seem to be aware of its existence, and can therefore take steps to overcome it.

In Experiment 2 we extended our examination of its effects on real-world educational materials to investigate another type of integration, which may afford protection against RIF. This integration is that provided by the form of presentation of the material itself; coherent text may, by virtue of its continuity, provide more protection from RIF than a series of unrelated statements. Reading sentences in an ordered or a disordered fashion influences the degree to which the material is integrated in memory (Carroll & Korukina, 1999; Shaddock & Carroll, 1997). Experiment 1 showed that coherently presented text is subject to RIF, but did not investigate whether the same text presented as a series of disordered statements is *even more* subject to RIF. There are educational applications in understanding this. For instance, when studying for an exam, students may vary the degree to which they integrate learned material. For some students, a learning strategy that entails summarising material into bullet points that capture separate elements of the information may suffice; for other students, material may need to be integrated into an overarching conceptual framework (commonly known as "deep" learning). As stated above, integrating material during the learning phase has been demonstrated as a factor that can protect $Rp-$ items from inhibition (Anderson & Bell, 2001; Anderson & McCulloch, 1999). With instructions that encouraged participants to either rehearse each item with previously studied items from the same category or to simply study the relationship between each category-exemplar word pair (such as fruit–orange), Anderson and McCulloch (1999) found that encouraging the integration of exemplars reduced RIF. In the classroom, the analogous situation is the presentation of material to be learned in an ordered (textually coherent) or a disordered fashion.

Experiment 2 aims to investigate both integration and another variable, type of test format, to determine the relevance of RIF to real-world learning, particularly in exam situations. Initial investigations failed to report RIF with recognition tests (Anderson & Bjork, 1994; Butler, Williams, Zacks, & Maki, 2001). Butler et al. (2001) observed RIF with a category-cued recall test, but not with a fragment-completion test, a word-fragment-cued recall test, a category-plus-fragment-cued recall test, and a category-plus-stem-cued recall test. It may be that item-specific cues, such as word fragments, release the inhibition impairing that item's recall (Butler et al., 2001). These conclusions have since proven to be inconclusive (Anderson, 2003), with other studies demonstrating RIF on tests of item recognition (Hicks

& Starns, 2004; Verde, 2004). Clearly, there is a need for a systematic investigation of the susceptibility to RIF of variations of test format typically found within educational settings, such as multiple choice, short answer, and essay-style tests. An additional departure from the standard RIF paradigm is that in real-world study settings, and in our study, the questions, multiple choice distractors, and required targets may not exactly repeat the wording of the material learned in the study phase. If RIF were observed with such materials, it would further strengthen the notion that memory inhibition is cue independent.

Experiment 1 found that JOL predictions are influenced by the memory inhibition of unpractised material, suggesting a degree of awareness about the results of inhibition. Judgements of learning were lower for Rp− items than for NRp items for those participants who demonstrated RIF effects. A review of a number of studies and theoretical discussions provides inconclusive evidence as to whether recall predictions can distinguish between those items that have and have not been inhibited in memory (Benjamin, Bjork, & Schwartz, 1998; Carroll & Shanahan, 1997). The addition of a judgement of learning before testing, but after retrieval practice, is included in Experiment 2 to determine whether the apparent awareness of inhibition is a reliable finding.

Thus, Experiment 2 aims to (1) examine the integration-protection hypothesis by manipulating type of material—integrated or not integrated; (2) determine whether RIF is apparent with real-world material that is tested in a variety of classroom test formats; and (3) replicate the finding of Experiment 1, that, where RIF is apparent, individuals can predict the impairment in Rp− items before the testing takes place.

Method

Participants

Forty-five undergraduate Psychology students with a mean age of 21.96 years ($SD = 6.28$, range of 18–47 years) were tested individually.

Design

The design was a 3 (retrieval-practice condition of learned information: Rp+, Rp−, and NRp) × 2 (text integration: ordered and disordered) × 3 (test format: multiple choice, short answer, and essay) mixed model. Retrieval practice and text integration were manipulated within participants. Test format was manipulated between participants and provided three experimental groups, with repeated measures on text integration creating a total of six experimental conditions, with 15 participants in each condition.

The inclusion of a within-participants test of the integrated text variable in the present study follows the suggestions of Carroll and Nelson (1993) who showed that such designs are more sensitive in detecting the effect of an independent variable on metacognitive judgements.

The dependent measures were (1) percentage correct recall and (2) magnitude of Judgement of Learning (JOL) ratings. These measures were calculated across each of the six conditions.

Materials

Four different texts were adapted from introductory psychology text-books; care was taken to ensure that each did not overlap considerably with general knowledge. The selected topics were Electroconvulsive Shock Therapy (Text A), Mental Disorders (Text B), Self-Actualisation (Text C), and Defining Intelligence (Text D). Each heading was centred at the top of the page as the text "category". Each text was approximately 500 words and an A4 page in length. For the disordered condition, the sentences from the original texts were arranged in alphabetical order, and those beginning with a pronoun were expanded to state the replaced noun.

Retrieval-practice materials. Two sets of retrieval-practice materials, each containing five statements, one or two sentences long, were created for each of the four texts, with a "category" heading at the top of the page. Some of the statements for practice comprised information from across two sentences. Retrieval practice consisted of the participant completing the missing phrase in each statement in response to its first letter, and thus constituted a cued-recall test. For example, for the original sentence: *Together, the first four needs are referred to as deficiency needs,* practice consisted of supplying the missing word for the frame: *Together, the first four needs are referred to as d_____ needs.* Participants were required to successfully retrieve the missing phrase for each statement three times, in its entirety.

Test materials. A multiple choice, short answer, and an essay-style test was created for each of the four texts, with each test having its "category" heading at the top of the first page. These tests aimed to replicate those given in educational contexts, and differed in the amount of self-initiated retrieval required. The multiple choice test presented participants with one correct answer and three incorrect but plausible alternative responses, most of which were from the teaching instruction manual accompanying the textbook. Appendix B shows parallel versions of the same questions in different test formats.

The short answer questions were the same as the multiple choice questions; each required a single unique response, and no half marks were given for incomplete answers. Pilot testing revealed that all questions were of similar difficulty level.

The essay-style test asked participants to recall as much information as they could remember from the text they had just read. They could list the information in bullet-point form if preferred. In this case, the task was less like writing an essay, and more similar to preparing a plan for an essay-style exam response. Nevertheless, for convenience, it is referred to as the essay test condition.

JOL rating materials. JOL ratings were given for each of the 10 questions per text, after the rehearsal practice had occurred and before the test. Judgements were made on a 6-point rating scale, where 1 = "*I feel I have not learned the information and will not recall it at a test in five minutes*" and 6 = "*I feel I have learned the information well and will certainly recall it correctly in five minutes*". Importantly, JOLs were made to the question only, without the answer.

Tasks.   Four distractor tasks were used, two relating to verbal tasks, and two to numerical tasks.

Procedure

Individual testing sessions lasted from 60 to 70 min. Counterbalancing the order of presentation of the texts and test formats was achieved using a stratified random sample based upon the 24 combinations of text materials, A, B, C, and D. In the learning phase, participants read the first text for 5 min knowing that their recall (test unspecified) of the information would be assessed later in the study. To replicate real study conditions, they were allowed to make notes during the study, but they could not access these at test. After 5 min, they read a second text for the same purpose.

Retrieval practice commenced immediately after completion of the learning phase. Five statements from one of the two texts they had just read (the Rp+ items) were read aloud. Each had one missing word cued by the first letter, to be completed by participants. If they were unable to complete a word, immediate feedback was given on their performance, with the intention of having participants correctly complete each statement three times. All statements had to be answered correctly three times nonconsecutively. No participant required more than five attempts to reach this criterion. Following retrieval practice, participants performed a distractor task for 3 min and then provided JOL ratings for the 10 questions associated with each text read during the learning phase. These questions comprised the five Rp+ and five Rp− items from the rehearsed text, and the ten NRp items from the unrehearsed text. Only the questions, not the answers, were presented in the JOL ratings phase, in order to prevent a further retrieval practice opportunity.

A second 3-min task followed JOL ratings, after which participants were given separate 5-min tests on the two texts they had just read. There was no knowledge of type of test format until this point, to ensure that all participants learned the material in the same way.

Upon completion of the two tests, the procedure was repeated using two different texts, with these texts being different in format (ordered or disordered) to the first two.

Results

Dependent measures consisted of (1) proportions correct and (2) JOL rating magnitude for Rp+, Rp−, and NrP information. To test for the main results of interest, the inhibition effect, mixed-model repeated measures analyses of variance were carried out with the factors being 2 (text order: ordered vs. disordered) × 3 (test format: multiple choice, short answer, essay) × 2 (item type: Rp− vs. NRp).

Participants correctly completed 83.13% ($SD = 9.01$) of the missing words for each statement during retrieval practice across conditions. This figure is comparable to that of previous researchers (Anderson et al., 1994; Anderson, Green, & McCulloch, 2000; Anderson & McCulloch, 1999; Dunn & Spellman, 2003; Macrae & McLeod, 1999) and indicates the success rate of retrieval practice in the present study.

Proportion correct.   The mean recall performance of participants according to item type, test format, and text order conditions are presented in Table 4. There was a significant effect of text order, $F(1, 42) = 4.93$, $p < .05$, $Mse = 521$, with ordered text being better recalled (mean = 0.43, $SEm = 2.53$) than disordered text (mean = 0.36, $SEm = 2.70$). There was also a significant inhibition effect, $F(1, 42) = 10.9$, $p < .01$, $Mse = 309$, with the mean performance on Rp− items (0.35, $SEm = 2.7$) poorer than that of NRp items (mean = 0.44, $SEm = 1.97$). The test format effect was also significant, $F(2, 42) = 27.94$, $p < .001$, $Mse = 717$, with the mean for multiple choice format (0.59) being significantly greater than that of short answer (0.36) and essay (0.23) formats. The only significant interaction was that between inhibition and test format, $F(2, 42) = 4.56$, $p < .01$, $Mse = 309$. Pairwise comparisons revealed no difference between Rp− and NrP item recall in the multiple choice condition, indicating no RIF effect with this type of test format. In the short answer condition, Rp− recall ($M = 30.67$,

TABLE 4
Mean proportions of correct and standard deviations by test format, and order of text, for Rp+, Rp−, and NRp items, Experiment 2

Type of test	Ordered text			Disordered text		
	Rp +	Rp −	NRp	Rp +	Rp −	NRp
Multiple choice	.77	.62	.63	.72	.60	.54
	(.82)	(.58)	(.50)	(.80)	(.66)	(.45)
Short answer	.69	.40	.50	.64	.21	.35
	(.82)	(.58)	(.50)	(.80)	(.66)	(.45)
Essay	.52	.14	.33	.44	.17	.28
	(.82)	(.58))	(.50)	(.80)	(.66)	(.45)

$SE = 4.52$) was significantly lower ($p < .05$) than NrP recall ($M = 43$, $SE = 3$), and in the essay condition, Rp− recall ($M = 15.33$, $SE = 2.91$) was also significantly lower ($p < .001$) than NrP recall ($M = 31.33$, $SE = 3.33$). Together, these analyses indicate the operation of RIF with the short answer and essay tests, but not in the multiple choice condition.

Judgements of learning.   Mean JOL ratings according to item type, test format, and text integration conditions are presented in Table 5. A 2 (text order) × 2 (item type: Rp− vs. NRp) × 3 (test format) repeated measures ANOVA on JOL magnitude revealed only one significant main effect: that of text order, $F(1, 42) = 14.23$, $p < .001$, $Mse = 0.81$. Ordered text was judged to be more likely to be remembered (mean = 3.6, $SEm = 0.12$) than disordered text (mean = 3.11, $SEm = 0.12$). No interactions were significant.

As in Experiment 1, then, participants were not able to predict the significant inhibition that occurred; that is, they did not know that Rp− items would be answered less accurately than NRp items.

TABLE 5
Mean proportion of JOL ratings and standard errors by test format, and order of text, for Rp+, Rp −, and NRp items, Experiment 2

Type of test	Ordered text			Disordered text		
	Rp +	Rp −	NRp	Rp +	Rp −	NRp
Multiple choice	.378	.357	.376	.369	.329	.33
	(.29)	(.26)	(.25)	(.33)	(.26)	(.25)
Short answer	.432	.357	.402	.384	.299	.304
	(.29)	(.26)	(.25)	(.33)	(.26)	(.25)
Essay	.474	.346	.334	.393	.298	.301
	(.29)	(.26)	(.25)	(.33)	(.26)	(.25)

As in Experiment 1, a further analysis was conducted separately for those participants who showed RIF and those who did not show RIF. As Table 6 shows, the mean JOL scores for those participants who showed RIF mirrored the proportion correct scores.

It would be desirable to be able to compare $Rp-$ and NRp conditions for each test format for participants who experienced exactly the same combinations of test questions. However, due to the small numbers, it was not possible to make the comparisons having regard to the counterbalancing conditions. Thus, it is possible that the split into high and low levels of inhibition was due to combinations of items that were particularly difficult or easy. Thus, any conclusions about the awareness of RIF must be regarded as tentative. When individual paired t-tests were performed between $Rp-$ and NRp items for those participants who showed RIF, there were no significant effects in any of the three test format conditions, due, no doubt, to the small sample sizes.

These results are promising but not firm (for the reasons given above) evidence that people are able to monitor the products of the inhibition effect. Further specifically designed studies with larger sample sizes are needed.

TABLE 6

Mean proportion of JOL predictions according to the presence or absence of RIF across test format conditions, Experiment 2

		$Rp+$		$Rp-$		NRp	
	N	Mean proportion correct	Standard error	Mean proportion correct	Standard error	Mean proportion correct	Standard error
RIF participants							
Multiple choice	8	.388	.042	.343	.043	.399	.032
Short answer	14	.400	.029	.308	.036	.322	.027
Essay	18	.450	.031	.304	.024	.353	.025
Non-RIF participants							
Multiple choice	22	.369	.027	.343	.019	.341	.019
Short answer	16	.415	.025	.346	.024	.380	.027
Essay	12	.410	.042	.350	.026	.265	.023

Results have been collapsed across text order, creating a total sample of 90 observations from 45 participants.

Discussion

Experiment 2 had three aims: to compare the operation of memory inhibition in classroom-like learning situations under conditions where the material itself was presented in an integrated or an isolated format; to investigate whether RIF occurs with different test formats; and to determine whether individuals can monitor the inhibited information. This study demonstrated that the degree of integration of the text did not provide the degree of protection against RIF that was observed in the experts in Experiment 1, suggesting that, at least for newly learned material, it is irrelevant whether the information presented is integrated or not.. The results also showed that RIF occurred in both short answer and essay tests, but not in the multiple choice test; and that individuals who experience RIF may perhaps be able to monitor the effect of that inhibition.

Previously, Anderson and colleagues (Anderson & Bell, 2001; Anderson & McCulloch, 1999) have demonstrated that encouraging the integration of learned category-exemplar word pairs reduces RIF. The idea here is that integration increases the interconnections between items and helps to transfer the benefits of retrieval practice for Rp+ items to the unrehearsed Rp− items. In the Anderson studies, participants were instructed to form interconnections based upon similarities and differences between word pairs, a manipulation not explicitly required in our study, although it would be expected this would occur automatically in integrated text.

Some studies that provide an alternative explanation for RIF—the strategy disruption hypothesis—would suggest that ordered text should be more susceptible to RIF than disordered text. According to the strategy disruption hypothesis (Basden & Basden, 1995; MacLeod, Dodd, Sheard, Wilson, & Bibi, 2003) RIF may occur because retrieval practice disrupts the original organisation of the learned information in memory. While Rp+ items benefit from retrieval practice, Rp− items become more difficult to remember. NrP items are easier to recall than the Rp− items because of a lack of practice and disruption (MacLeod et al., 2003). Recalling information that has been learned in a disordered and isolated fashion may not be susceptible to the disruption resulting from retrieval practice. Hence, RIF may be reduced for a disordered text. Indeed in our study, the Rp− versus NRp difference *was* greater (though not significantly so) in both the ordered short answer condition and in the ordered essay condition than in the disordered versions. Despite the plausibility of the strategy disruption hypothesis, an explicit test of this hypothesis is yet to be undertaken in reference to RIF.

With inconsistencies in the literature surrounding the operation of RIF with recall versus recognition testing (Anderson & Bjork, 1994; Butler et al., 2001; Veiling & van Knippenberg, 2004; Verde, 2004), no clear prediction was made concerning RIF's interaction with test format. Our results indicated an absence of RIF for the multiple choice test in both order conditions. In the short answer and essay tests, however, RIF was clearly apparent. Essay and short answer tests differ from multiple choice tests in the degree of self-initiated recall required for correct answers: In multiple choice tests, selecting the correct response is more a process of recognition (Oosterhof, 2003) than retrieval. A failure to observe RIF with recognition, but not a recall, test is consistent with the results of Butler et al. (2001) and Koustaal, Schacter, Johnson, and Galluccio (1999); the latter argued that the retrieval cues are enough to overcome the suppression of the nonreviewed Rp− items, with such cues reexposing the individual to the initial learning episode. Such an explanation suggests that inhibition is a flexible process that can be overcome with item-specific cues. Other studies *have* found RIF with tests of item recognition when the dependent measure was reaction time latencies (Veiling & van Knippenberg, 2004). Observing RIF with reaction time latencies (Veiling & van Knippenberg, 2004) and not with correct response rates in recognition tests (Butler et al., 2001; Koustaal et al., 1999) suggests that Rp− items are inhibited before the completion of a recognition test. The inclusion of reaction time latencies and confidence ratings (for accuracy of lures and targets) within the retrieval-practice paradigm may clarify this possibility. In addition, these measures may also indicate the extent to which RIF was masked by guessing in the present study, given that multiple choice tests are somewhat susceptible to guessing (Oosterhof, 2003).

It may also be the case that our results were confounded by the choice of distractors used in Experiment 2. These distractors were not items that had been exposed in the text. An alternative way of doing the multiple choice test would be to include distractors that were studied earlier, as is often the case in real educational settings. Perhaps our Rp− items in multiple choice were indeed inhibited, but not to the point where they were indistinguishable from completely new items. The familiarity of the target answers among the new distractors in our study may have been responsible for the apparent absence of RIF. Some caution should be exercised, then, in concluding that multiple choice tests are not affected by RIF, especially in light of the studies mentioned above (Hicks & Starns, 2004; Verde, 2004) that RIF has been observed in recognition memory tests.

The other line of enquiry in Experiment 2 concerned metamemorial monitoring of the inhibition effect. The failure of participants to accurately detect Rp− inhibition was at odds with the finding of Experiment 1, where participants accurately monitored the Rp− versus NRp difference. However, when participant data were analysed according to whether or not an individual showed RIF, the findings were generally consistent with Experiment 1. The impracticability of comparing subjects who received exactly the same combinations of items in the RIF and non-RIF groups tempers any conclusions we are able to draw, but if RIF monitoring is possible, then there are implications for classroom learning, such as that students studying for an exam can assess their poor performance on some items with some reliability. In considering exactly what the participants are monitoring, we are indebted to a reviewer (B. Levy, personal communication) who suggests that what people are tapping into is the results or products of the inhibition, rather than the inhibition itself. Awareness of the process of inhibition is not necessary: They simply register that the Rp− item, for whatever reason, seems harder to recall than other items.

Our results were consistent with previous work on monitoring of integrated text. Previous research has demonstrated that JOLs are sensitive to both the amount of learning (Shaddock & Carroll, 1997) and text coherence and context (Carroll & Korukina, 1999; Mazzoni & Nelson, 1995). In both instances, individuals predict greater recall for overlearned items and for material learned in an ordered sequence. In the present Experiment 2, the integrated text did indeed receive significantly higher JOL ratings than the nonintegrated text.

GENERAL DISCUSSION

We have established that RIF is a robust phenomenon that occurs with real-world classroom material standard in formal learning situations. It is apparent that material of this type is not immune to the detrimental effects of retrieval practice. We have also established a boundary condition on the protection that integrated material affords against RIF. In Experiment 1, integration of material that was provided by expertise in the domain did indeed provide this protection. The background knowledge of the experts effectively meant that the newly learned material was not inhibited in the same way as it was for novices. This accords with Anderson and McCulloch's (1999) observation that experts are better able to recall individual items from within a comparatively larger body of information. In Experiment 2, a different method of inducing "integration" of the material was manipulated, one that relied on presentation format, rather than individual expertise, as a means of inducing integration. This involved comparing expository prose format, which itself may encourage spontaneous integration of the material, with a series of disparate and apparently unconnected facts. Experiment 2 found that presenting information to students in a holistic way, which might better encourage integration, did not protect them from RIF.

It appears that individuals can indeed monitor the effect RIF will have on their later test performance, at least in short answer and essay tests.

Experiment 1 also supports previous studies that have found that inhibitory effects reduce dramatically over the course of 24 hours. In terms of real-world applications for this data, this study suggests that RIF is not an issue for long-term learning, being certainly a transitory effect. Furthermore, in common tertiary-level learning scenarios where material is presented and learned in a holistic and integrated manner, RIF is not likely to be an issue. However, in some educational contexts, it is conceivable that new and unconnected material may be learned, practised, and finally quizzed within a single learning session. Such a teaching practice would be highly inadvisable based upon these findings, as it would almost certainly result in significant levels of RIF.

Despite pressures from teachers and parents to effectively practise and review material in the lead-up to an exam, an individual's exam performance may be impaired by this process. In some circumstances an individual's exam performance may be better off without excessive overlearning (see also Nelson & Leonesio, 1988) of some items at the expense of others. As suggested by Macrae and MacLeod (1999), retrieval practice may indeed be detrimental to exam performance, particularly when the test format is short answer or essay. However, where retrieval practice of a subset of items is encouraged as a study technique, being aware of the likely outcomes of retrieval practice may allow students to compensate in some way prior to a test.

REFERENCES

Anderson, M. C. (2003). Rethinking interference theory: Executive control and the mechanisms of forgetting. *Journal of Memory and Language, 49*, 415–445.

Anderson, M. C., & Bell, T. (2001). Forgetting our facts: The role of inhibitory processes in the loss of propositional knowledge. *Journal of Experimental Psychology: General, 130*(3), 544–570.

Anderson, M. C., Bjork, E. L., & Bjork, R. A. (2000). RIF: Evidence for a recall-specific mechanism. *Psychonomic Bulletin and Review, 7*(3), 522–530.

Anderson, M. C., & Bjork, R. A. (1994). Mechanisms of inhibition in long-term memory: A new taxonomy. In D. Dagenbach & T. Carr (Eds.), *Inhibitory processes in attention, memory and language* (pp. 265–326). San Diego, CA: Academic Press.

Anderson, M. C., Bjork, R. A., & Bjork, E. L. (1994). Remembering can cause forgetting: Retrieval dynamics in long-term memory. *Journal of Experimental Psychology: Learning, Memory, and Cognition, 20*(5), 1063–1087.

Anderson, M. C., Green, C., & McCulloch, K. C. (2000). Similarity and inhibition in long-term memory: Evidence for a two-factor theory. *Journal of Experimental Psychology: Learning, Memory, and Cognition, 26*(5), 1141–1159.

Anderson, M. C., & McCulloch, K. C. (1999). Integration as a general boundary condition on RIF. *Journal of Experimental Psychology: Learning, Memory, and Cognition, 25*(3), 608–629.

Anderson, M. C., & Spellman, B. A. (1995). On the status of inhibitory mechanisms in cognition: Memory retrieval as a model case. *Psychological Review, 120*(1), 68–100.

Barnier, A. J., Hung, L., & Conway, M. A. (2004). RIF of emotional and unemotional autobiographical memories. *Cognition and Emotion, 18*(4), 457–477.

Basden, D. R., & Basden, B. H. (1995). Some tests of the strategy disruption interpretation of part-list cuing inhibition. *Journal of Experimental Psychology: Learning, Memory, and Cognition, 21,* 1656–1669.

Bäuml, K.-H. (2002). Semantic generation can cause episodic forgetting. *Psychological Science, 13,* 356–360.

Bäuml, K.-H., & Kuhbandner, C. (2003). RIF and part-list cuing in associatively structured lists. *Memory and Cognition, 31*(8), 1188–1197.

Benjamin, A. S., Bjork, R. A., & Schwartz, B. L. (1998). The mismeasure of memory: When retrieval fluency is misleading as a metamnemonic index. *Journal of Experimental Psychology: General, 127*(1), 55–68.

Butler, K. M., Williams, C. C., Zacks, R. T., & Maki, R. H. (2001). A limit on RIF. *Journal of Experimental Psychology: Learning, Memory, and Cognition, 27*(5), 1314–1319.

Carroll, M., & Korukina, S. (1999). The effect of text coherence and modality on metamemory judgements. *Memory, 7*(3), 309–322.

Carroll, M., & Nelson, T. O. (1993). Effect of overlearning on the feeling of knowing is more detectable in within-subjects than in between-subjects designs. *American Journal of Psychology, 106,* 227–235.

Carroll, M., & Shanahan, C. (1997). The effect of context and metamemory judgements on automatic processes in memory. *Acta Psychologica, 97,* 219–234.

Dunn, E. W., & Spellman, B. A. (2003). Forgetting by remembering: Stereotype inhibition through rehearsal of alternative aspects of identity. *Journal of Experimental Social Psychology, 39,* 420–433.

Hicks, J. L., & Starns, J. J. (2004). RIF occurs in tests of item recognition. *Psychonomic Bulletin and Review, 11*(1), 125–130.

Johnson, S. K., & Anderson, M. C. (2004). The role of inhibitory control in forgetting semantic knowledge. *Psychological Science, 15*(7), 448–153.

Koustaal, W., Schacter, D. L., Johnson, M. K., & Galluccio, L. (1999). Facilitation and impairment of event memory produced by photograph review. *Memory and Cognition, 27*(3), 478–493.

Levy, B. J., & Anderson, M. C. (2002). Inhibitory processes and the control of memory retrieval. *Trends in Cognitive Science, 6*(7), 299–305.

MacLeod, M. D. (2002). RIF in eyewitness memory: Forgetting as a consequence of remembering. *Applied Cognitive Psychology, 16,* 135–149.

MacLeod, C. M., Dodd, M. D., Sheard, E. D., Wilson, D. E., & Bibi, U. (2003). In opposition to inhibition. *Psychology of Learning and Motivation, 43,* 163–214.

MacLeod, M. D., & Macrae, C. N. (2001). Gone but not forgotten: The transient nature of RIF. *Psychological Science, 12*(2), 148–152.

Macrae, C. N., & MacLeod, M. D. (1999). On recollections lost: When practice makes imperfect. *Journal of Personality and Social Psychology, 77*(3), 463–473.

Mazzoni, G., & Nelson, T. O. (1995). Judgments of learning are affected by the kind of encoding in ways that cannot be attributed to the level of recall. *Journal of Experimental Psychology: Learning, Memory and Cognition, 21,* 1263–1274.

Nelson, T. O., & Leonesio, R. J. (1988). Allocation of self-paced study time and the "labour-in-vain effect". *Journal of Experimental Psychology: Learning, Memory, and Cognition, 14*(4), 676–686.

Nelson, T. O., & Narens, L. (1994). Why investigate metacognition? In J. Metcalf & A. P. Shimamura (Eds.), *Metacognition: Knowing about knowing* (pp. 1–25). Cambridge, MA: MIT Press.

Oosterhof, A. (2003). *Developing and using classroom assessments* (3rd edn). New York: Pearson Education.

Pelegrina, S., Bajo, M. T., & Justicia, F. (2000). Differential allocation of study time: Incomplete compensation for the difficulty of the materials. *Memory, 8*(6), 377–392.

Perfect, T. J., Moulin, C. J., Conway, M. A., & Perry, E. (2002). Assessing the inhibitory account of RIF with implicit memory tests. *Journal of Experimental Psychology: Learning, Memory, and Cognition, 28*(6), 1111–1119.

Radvansky, G. A., & Zacks, R. T. (1991). Mental models and the fan effect. *Journal of Experimental Psychology: Learning, Memory and Cognition, 17*, 940–953.

Roediger, H. L., & Karpicke, J. D. (2006). Test-enhanced learning: Taking memory tests improves long-term retention. *Psychological Science, 17*, 249–255.

Saunders, J., & MacLeod, M. D. (2002). New evidence on the suggestibility of memory: The role of RIF in misinformation effects. *Journal of Experimental Psychology: Applied, 8*(2), 127–142.

Schwartz, B. L., & Perfect, T. J. (2002). Introduction: Toward an applied metacognition. In T. J. Perfect & B. L. Schwartz (Eds.), *Applied metacognition* (pp. 1–14). Cambridge, UK: Cambridge University Press.

Shaddock, A., & Carroll, M. (1997). Influences on metamemory judgements. *Australian Journal of Psychology, 49*(1), 21–27.

Shaw, J. S., Bjork, R. A., & Handal, A. (1995). RIF in an eyewitness-memory paradigm. *Psychonomic Bulletin and Review, 2*(2), 249–253.

Shivde, G., & Anderson, M. C. (2001). The role in inhibition in meaning selection: Insights from retrieval inducted forgetting. In D. Gorfein (Ed.), *On the consequences of meaning selection: Perspectives on resolving lexical ambiguity* (pp. 175–190). Washington, DC: American Psychological Association.

Smith, E. E., Adams, N., & Schorr, D. (1978). Fact retrieval and the paradox of interference. *Cognitive Psychology, 10*, 438–464.

Veling, H., & van Knippenberg, A. (2004). Remembering can cause inhibition: Retrieval-induced inhibition as cue independent process. *Journal of Experimental Psychology: Learning, Memory, and Cognition, 30*(2), 315–318.

Verde, M. F. (2004). The retrieval practice effect in associative recognition. *Memory and Cognition, 32*(8), 1265–1272.

APPENDIX A

1. What did teachers first notice was odd about David's behaviour? Talked to his dead uncle.
2. Why was David unable to finish high school or obtain a job? He was agitated and verbally aggressive.
3. Which of David's delusions is a particular form of delusion known as Capgras syndrome? His belief that his aunt had been replaced by a double.
4. How could David's conversational speech be described? Highly disorganised.
5. David's tendency to jump from topic to topic and talk illogically is known by psychologists as? Associative splitting.
6. How did David demonstrate inappropriate affect? He laughed when thinking about the death of his uncle.
7. What was David's most common form of auditory hallucination? Hearing his dead uncle.

APPENDIX B

Examples of alternative forms of the same questions in Experiment 2.

Text: Electroconvulsive Shock Therapy

Extract from the text:

"When ECT is applied in the traditional manner bilaterally (with the current running across both of the brain's hemispheres), the patient typically loses memory for events that occurred a day or two before the treatment."

Multiple choice

What type of memory loss may be associated with bilateral ECT?

A. Events that occurred the day of treatment
B. Pictorial memories
C. Events occurring a day or two before treatment
D. Verbal memories

Short answer

What type of memory loss may be associated with bilateral ECT?

Free recall

Please recall as many points that you can remember and learned about when reading about Electroconvulsive Shock Therapy. You may list your points in bullet-point form on this sheet of paper (both sides if required).

Note: Retrieval practice for this item was of the following form:

When ECT is applied in the traditional manner bilaterally (with the current running across both of the brain's hemispheres), the patient typically loses memory for e_____ that occurred a d___ o__ t___ before the treatment.

EUROPEAN JOURNAL OF COGNITIVE PSYCHOLOGY
2007, 19 (4/5), 607–627

The effects of selective attention on perceptual priming and explicit recognition in children with attention deficit and normal children

Soledad Ballesteros

Department of Basic Psychology II, Universidad Nacional de Educación a Distancia, Madrid, Spain

José M. Reales

Department of Methodology of Behavioural Sciences, Universidad Nacional de Educación a Distancia, Madrid, Spain

Beatriz García

Department of Basic Psychology II, Universidad Nacional de Educación a Distancia, Madrid, Spain

Perceptual priming and recognition for attended and unattended pictures at encoding, compared to nonstudied pictures were examined in second and fifth grade schoolchildren with attention deficit (AD) and children without AD. In the study, a visual perceptual priming paradigm was combined with a selective attention procedure at encoding to look for the influence of attention in implicit and explicit memory tasks. The findings showed preserved perceptual priming for attended objects at encoding in second and fifth grade AD children and normal children but only the older children showed reduced perceptual priming for unattended pictures, a result that has been reported in adults (Ballesteros, Reales, García, & Carrasco, 2006). Overall, AD children performed more poorly in the picture fragment completion task than control children, exhibiting a general deficit in the task. The findings suggest that substantial developmental changes occurred in both groups

Correspondence should be addressed to Soledad Ballesteros, Dept. Psicología Básica II, Facultad de Psicología, UNED, Juan del Rosal, 10, 28040 Madrid, Spain. E-mail: mballesteros@psi.uned.es

Preliminary results were presented at the Third International Conference on Memory, Valencia, Spain, July 2001. We acknowledge the financial support of the *Comunidad de Madrid* (Grant Ref. 0012–00). The authors wish to thank Paloma Gómez for helping with data collection and to the children, parents, teachers, and school directors whose cooperation made the present study possible. We are very grateful to Morton Heller and Ashley Clark for their helpful comments in a previous version of this paper. We also thank Lisa K. Son, Veronica Dark, and two anonymous reviewers for their thoughtful reviews that helped us to improve the manuscript.

and that attention does not dissociate performance in implicit and explicit memory tasks. The preserved perceptual priming and recognition observed in AD children indicate that they performed normally in effortless memory tasks in which stimuli remain present during testing. These results may have important practical implications as these preserved abilities may be used in the rehabilitation of these children.

Children's difficulty in attention deployment and impulsivity can increase the risk for persisting conduct problems (Snyder, Prichard, Schrepferman, Patrick, & Stoolmiller, 2004) and memory deficits during complex memory tasks (Cornoldi, Barbieri, Gaiani, & Zocchi, 1999). At a clinical level, attention-deficit hyperactivity disorder (AD) is a prevalent childhood psychological disorder (Barkley, 1997; Biederman, 1998) and is the diagnosis label used for children with significant problems of attention and impulsiveness (American Psychiatric Association, 1994). Today, this disorder is one of the most common developmental disabilities in children and typically manifests prior to the age of 7 years affecting between 3% and 7% of the school-aged population. Moreover, it is three times more frequent in boys than in girls (Barkley, 1998). These children have lower academic achievement, higher risk of not being promoted to the next academic grade, behavioural problems, lack of work concentration, and difficulty in controlling impulses (Weiss & Hechtman, 1993).

The main behavioural assessment techniques used in the diagnosis of this disorder include psychometric tests and rating scales completed by parents and teachers of these children (Mitchell, Chavez, Baker, Guzman, & Azen, 1990). Only recently, some experimental paradigms have been used to study the cognitive processes of attention deficit (AD) children. One stream of cognitive studies is related to the covert attention paradigm developed by Posner (1980). The covert attentional system allows attention to be directed to certain regions of visual space without eye movements. This ensures a more efficient processing of stimuli located in the attended area of space. Using the covert attention paradigm, it has been found that children with AD and hyperactivity show a strong anchorage of attention upon a cued location and an inability to shift covert attention to another location in the visual field (McDonald, Bennett, Chambers, & Castiello, 1999). Furthermore, these children but not control children of the same age displayed an attentional bias towards both positively and negatively valenced cues (Sonuga-Barke, de Houwer, de Ruiter, Ajzenstzen, & Holland, 2004).

Although previous research has focused on the experimental dissociations between implicit and explicit memory tasks in different developmental and clinical settings, very few studies have been conducted on children with AD (Ballesteros & Reales, 2004; Parkin, & Streete, 1988). To our knowledge, only two other recent studies have examined the possible dissociations between implicit and explicit memory tasks in AD children (Aloise, McKone, & Heubeck, 2004; Burden & Mitchell, 2005). However, as far as we know, this is the first study conducted to evaluate the effects of selective attention at encoding on implicit and explicit memory tests in AD children and control children at two developmental levels.

The present study examines the effect of selective attention in a implicit memory task (picture fragment completion task; PFCT) and in an explicit memory task (visual recognition) performed by schoolchildren with AD and control children of second and fifth school grades. Implicit and explicit memory refer to two different ways of accessing previously encoded information.

EXPLICIT AND IMPLICIT MEMORY

Explicit memory requires the conscious retrieval of previous experience and it is usually assessed by recognition and recall tests. In contrast, implicit memory refers to facilitation with previously encountered stimuli and it is assessed by indirect tests with no reference to previous experience (Schacter, 1987). Implicit memory is demonstrated by showing perceptual (repetition) priming; that is, better performance with previously encountered stimuli compared to new stimuli. Early studies on implicit memory focused mainly on verbal materials. More recently, researchers have focused on nonverbal visual stimuli (Biederman & Cooper, 1992; Schacter, Cooper, & Delaney, 1990), and 3-D objects presented to touch (Ballesteros, Reales, & Manga, 1999; Reales & Ballesteros, 1999). Perceptual implicit memory relies on processing physical characteristics of the studied stimuli. Perceptual priming, however, is not sensitive to all the perceptual characteristics of the stimuli. For example, visual studies with adults have shown significant perceptual priming for pictures of familiar (Biederman & Cooper, 1992) and unfamiliar objects (Cooper, Schacter, Ballesteros, & Moore, 1992) when the size and orientation of the stimuli changed from study to test. In contrast to perceptual implicit memory, conceptual implicit memory requires partici-pants to produce the studied items in response to test cues that are semantically or conceptually related to the studied stimuli. In this paper, we refer exclusively to perceptual implicit memory.

THE ROLE OF ATTENTION IN IMPLICIT AND EXPLICIT MEMORY

It is well documented that attention is required for the formation of enduring memory traces when previously encoded information is consciously, voluntarily recovered from explicit (episodic) memory (e.g., Craik, Govoni,

Naveh-Benjamin, & Anderson, 1996; Fisk & Schneider, 1984; Mulligan, 1998; Rock & Gutman, 1981). Adult studies using verbal material have shown attentional effects in both implicit and explicit memory tests (Crabb & Dark, 1999; MacDonald & MacLeod, 1998). However, the role of selective attention in implicit memory for pictures has not been widely studied. We have addressed this question in a series of studies conducted with adult participants (Ballesteros & Reales, 1998; Ballesteros & Reales, 2005; Ballesteros, Reales, García, & Carrasco, 2006). These studies have consistently shown that implicit memory for pictures requires attention at encoding. Stronger perceptual priming as well as better recall and recognition for attended than for unattended pictures have been reported. The main question we asked in the present study is whether selective attention plays a main role in the implicit memory of AD children.

THE PRESENT STUDY

In this study, we used a widely explored experimental paradigm to study implicit memory in children with AD and in normal children, and its possible dissociation from episodic memory. We combined a repetition priming paradigm with a selective attention procedure at encoding (Ballesteros, Reales, García, & Carrasco, 2006) to investigate the influence of attention in implicit and explicit memory processes of these children at two different school grades (second and fifth grades). The selection of these two age groups allowed us to study the developmental and functional implications of the ability to integrate spatially disparate elements into cohesive whole pictures (Dukette & Stiles, 1996, 2001; Kimchi, Hadad, Behrmann, & Palmer, 2005). As has been recently shown, the selection of an appropriate age seems important for the detection of group differences between children with attentional difficulties and control children (Drechsler, Brandeis, Földényi, Imhof, & Steimhausen, 2005). The question of whether AD children differ from normal children in cognitive processes is an underinvestigated area. In this study, we report an innovative use of adult experimental paradigms to examine age and individual differences in children's performance, trying to bridge the gap between adult and child cognitive research.

Perceptual priming as a measure of implicit memory is a robust phenomenon. Age invariance usually exists in implicit memory tasks for pictures (e.g., Mitchell, 1989), haptic objects (Ballesteros & Reales, 2004), edible and nonedible odours (Fusari & Ballesteros, 2006), and auditory events (Ballesteros, González, et al., in press), not only in normal ageing but also in Alzheimer's disease patients for different stimuli such as 3-D objects

(Ballesteros & Reales, 2004), pictures (Ballesteros, Reales, & Mayas, 2007), and words (Park et al., 1998). The robustness of perceptual priming allows us to predict that implicit memory would be normal in AD children and possibly it would not differ from control children of the same age. Moreover, recent neuroimaging studies suggest that visual priming depends on the integrity of extrastriate brain areas, which seem to function normally in AD children (this issue is taken up in the Discussion section). So, we hoped to find repetition priming for attended objects in these children.

To gather evidence of the relation between attention and implicit memory in AD and control children, we used the selective attention paradigm employed previously with adults (Ballesteros, Reales, García, & Carrasco, 2006). In this paradigm, we manipulated attention by using overlapping outline drawings of familiar objects in two different colours, one in blue and the other in green (Rock & Gutman, 1981). In the overlapping figures procedure, the participant's attention is directed at the figure that appears at a specific colour by asking the participant to name the object of that colour as quickly as possible. Implicit memory was tested using a PFCT, which involves presentation of degraded stimuli that become progressively clearer over time (Snodgrass, Smith, Feenan, & Corkin, 1987). Previous studies with normal children have used this task to assess implicit memory in young schoolchildren because it is a nonverbal task that enables the testing of children unable to read (e.g., Cycowicz, Friedman, Snodgrass, & Rothstein, 2000; DiGiulio, Seidenberg, O'Leary, & Raz, 1994; Lorsbach & Worman, 1989; Parkin & Streete, 1988; Russo, Nichelly, Gibertoni, & Cornia, 1995). Using PFCT, Cycowicz et al. (2000) reported a reliable developmental trend from 5-year-old children to young adult participants in both implicit (picture fragment completion) and explicit (free recall and recognition) performance. We used the same implicit task to gather information on the developmental changes occurring in the ability of AD children to integrate spatially distributed elements into cohesive pictures at two different ages (e.g., Dukette & Stiles, 1996). For explicit memory, we did not expect differences between AD groups and control groups as we selected to assess explicit memory a task with very low memory demands (see Burden & Mitchell, 2005, for a similar argument). Explicit recognition was assessed using an "old–new" recognition test. According to previous results in the episodic memory literature, we expected high recognition for attended pictures during encoding but little or no recognition at all for unattended pictures (Goldstein & Fink, 1981; Rock & Gutman, 1981).

METHOD

Participants

AD children were identified using multiple methods and informants (parents and teachers) and were assessed during the 6 months previous to the experimental study. See later for the indicators that defined the construct. The total initial sample consisted of 374 children, 7, 8, 10, and 11 years' old, attending the second and fifth grades at three different elementary schools located in the Madrid area (capital and suburbs). Informed consent was obtained from schoolmasters, teachers, and parents. From this initial sample, four groups of 12 children were selected following the procedure described below. Two groups corresponded to the second and fifth grade AD children. Two control groups were also selected from the same schools, each composed of 12 normal children. One second grade child and two fifth grade AD children did not complete the experiment and were excluded from further analysis: The second grade child was excluded due to technical problems and the two fifth grade children became ill and could not complete the experiment. The final AD groups consisted of 11 and 10 participants each. Children in the control groups attended the same schools and grades as the AD children.

To select these four experimental groups, a total of 374 children completed a selective and sustained attention test standardised for the Spanish population, the Thurstone and Yela Face Test (TYFT; Thurstone & Yela, 1968), in small groups at their schools. Children who scored lower than the 20th percentile in the TYFT were preselected as possible AD children. Then, the teachers and parents of these children were asked to fill out the ECI (Child Behaviour Assessment) questionnaire adapted to the Spanish population (Manga, Garrido, & Pérez-Solís, 1997). This questionnaire is based on the Child Behaviour Checklist (CBCL; Achenbach & Edelbrock, 1983) and was standardised with a sample of 1230 Spanish schoolchildren from 7 to 11 years old; the informants were the parents and teachers of those children. Children who scored at the 93rd percentile or higher in the ECI were selected as the AD groups and participated in the experimental session. The 93rd percentile was the cutoff as proposed by Manga et al. (1997). So, inclusion criteria for the AD groups were: (a) a score lower than the 20th percentile in the TYFT test, and (b) a score in the clinical range (in the 93rd percentile or higher) in the attention scale of the ECI questionnaire, filled in by the teachers and the parents of these children. Participants selected as control scored higher than the 50th percentile in the TYFT and lower than the 50th percentile in the ECI Scale. All participants reported normal or corrected-to-normal visual acuity and normal colour vision.

Stimuli and equipment

The stimuli were presented through a 14-inch colour monitor of a PC-compatible 486 computer, with a resolution of 640×480 pixels. One hundred and twenty line drawings were selected from the Snodgrass and Vanderwart (1980) picture set. The pictures were approximately 10×10 cm^2 subtending a visual angle of approximately 4×4 degrees. During the study phase, the pictures were depicted with a green or a blue outline (see Figure 1). In each trial, two overlapped pictures were presented at the screen, one in green and the other in blue.

At the test phase, the stimuli were fragmented outlines of the stimuli presented in black on a white background. The pictures were digitised and saved in graphic BMP format to be presented on a PC monitor. In order to fragment the pictures, a 40×30 grid made up of 16×16 pixel blocks was superimposed on the projected image. All of the 1200 blocks that contained some black pixels were identified (Snodgrass et al., 1987). The information was stored in an array and was randomly permuted. The deleted block rate of the image followed from the exponent function:

$$P = 0.7 \ e^{8.0}$$

Each picture was stored as a fragmented image at eight levels of completion (see Figure 2). Level 1 of fragmentation corresponded to the most fragmented image and Level 8 to the complete picture outline. The

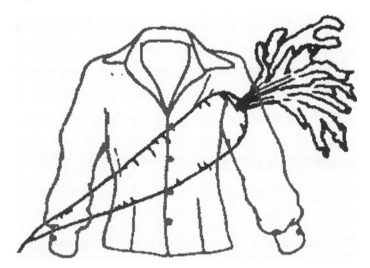

Figure 1. This diagram depicts the two overlapped pictures presented at the study phase of our study. The outline of one picture (the shirt) was green and the outline of the other picture (the carrot) was blue. The pictures were taken from Snodgrass and Vanderwart (1980). To view this figure in colour, see the online version of the Journal.

Figure 2. This diagram depicts the eight different levels of one of the fragmented pictures used in the picture fragment completion test. These fragmented pictures were prepared using the Snodgrass et al. (1987) fragmentation algorithm.

proportion deleted pixel blocks were 0.91, 0.88, 0.83, 0.76, 0.65, 0.51, 0.30, and 0.00 from Level 1 to Level 8, respectively.

Each child was tested individually in a quiet room at the school. He/she performed a cover study phase consisting of naming the attended (the green) picture as quickly as possible from two overlapping drawings (one green and the other blue).

After a 5-min filler task, implicit memory was assessed incidentally using PFCT. Finally, after completing the implicit test, participants performed an explicit "old–new" recognition task to assess explicit memory.

Experimental design

The main experimental design consisted of a 3 (study conditions: attended, unattended, and nonstudied pictures) × 2 (memory tests: implicit and explicit) × 2 (groups of children: AD and normal children) × 2 (grades: second grade and fifth grade) mixed factorial design. The first two variables were within subjects, whereas the last two variables were between subjects. In addition, 120 stimuli were randomly divided into three sets of 40 stimuli each. For counterbalancing, each of these sets appeared as attended (green), unattended (blue), or nonstudied pictures for four observers. These three sets were further subdivided into two subsets of 20 stimuli each that were presented equally often at the implicit and explicit tests. It should be noted that the particular colour of the attended or unattended picture is not important for object perception (see Rock & Gutman, 1981).

Study phase

The experiment started with a study phase. Upon entering the room, the child was seated directly in front of the computer screen at a distance of approximately 50 cm. In this phase, they were presented with 40 trials consisting of 40 green and 40 blue overlapping familiar objects, plus 5 trials at the beginning and 5 at the end to avoid primacy and recency effects. The encoding task was to name as quickly as possible the green (attended) object. At each trial, the child pressed a key when he/she was ready and, after 1500 ms, two overlapped outlines of familiar pictures appeared at the centre of the computer screen for 500 ms, one in blue and the other in green. After the oral response, the experimenter recorded the object's name using the computer's keyboard to assess later whether it was correct or incorrect.

After completing the study phase, all of the children performed a 5-min distractor task consisting of naming as many cartoon characters as possible during this time. The experimenter wrote down the names.

Test phase

When the 5-min distractor task was over, participants performed the PFCT (Snodgrass et al., 1987) incidentally. The presentation of the fragmented pictures followed the ascending method of limits. Children were presented with 60 progressively less fragmented (more completed) pictures (20 attended, 20 unattended, and 20 new) in a different random order for each participant. Each trial started at Level 1 of fragmentation, the more fragmented picture (see Figure 2, top left picture). Once the stimulus appeared at the screen, if the child did not respond in 1.5 s, the next more completed level was automatically shown (Level 2, then 3 ... until Level 8). The task consisted of identifying the picture as soon as possible, by pressing the spacebar of the keyboard as soon as they identified the object. The experimenter entered manually into the computer the participants' verbal responses stopping the sequence. The presentation order of the attended, unattended, and nonstudied stimuli was randomised for each participant. The child's performance was assessed by the fragmented level at which the pictures were correctly named, which was recorded by the computer.

After completing the implicit memory test, explicit memory was assessed by an "old–new" recognition test. Participants were presented with 60 pictures, 20 attended, 20 unattended, and 20 new pictures of familiar objects not presented in the implicit memory test. The pictures were presented one by one at the centre of the computer screen in a different random order for each participant and were asked to say "old" to the pictures presented during the study phase either in blue or in green colour. They were asked to respond "new" to pictures that were not presented at the study phase. Each picture remained on the screen until the participant responded or for a maximum of 5 s. The experimenter typed the responses on the keyboard, which were recorded by the computer program. Performance was assessed by the total number of correct attended pictures, correct unattended pictures, and correct rejections.

RESULTS

Results from the implicit and the explicit memory tasks were analysed separately. First, we report the results from the implicit memory test followed by the results obtained in the explicit recognition test.

Implicit memory

The average level of completion needed to identify the attended, unattended, and nonstudied pictures in the PFCT is shown in Figure 3 as a function of study condition, group, and school grade. Perceptual priming is shown when the average level of completion needed to identify the picture is lower for pictures presented at study compared to nonstudied pictures. In other words, the effect of implicit memory is shown by the difference between the attended/unattended identification thresholds and nonstudied fragmentation thresholds. The superior performance for attended pictures is shown by the lower threshold for picture identification (see Figure 3). The pictures were identified at a more fragmented level when children named (attended stimulus) the picture during the study phase than when they did not name the picture (unattended stimulus), or it was not shown at the study phase of the experiment (nonstudied picture). This result was confirmed by a mixed factorial ANOVA: 3 study conditions × 2 groups of children × 2 grades on the identification thresholds. The first variable was a within-subject variable, whereas the last two where between-subject variables. The main effect of study condition was significant, $F(2, 82) = 29.50$, $MSI = 0.116$, $p < .001$. *Post hoc* tests indicated that the identification threshold was lower for attended than for nonstudied pictures (mean threshold 5.43 vs. 5.95, respectively; $p < .01$) but the identification threshold for unattended pictures did not differ from nonstudied pictures (mean 5.85 vs. 5.95, respectively; $p > .1$). Attended pictures at study were identified at a more fragmented level than nonstudied stimuli showing significant priming effects but no priming was found for unattended objects.

Group was highly significant, $F(1, 41) = 8.11$, $MSE = 0.173$, $p < .007$. AD children identified the fragmented pictures at a more completed level (higher threshold) than control children. In other words, AD children needed to see more of the stimulus than control did before being able to name it (5.93 vs. 5.57, respectively).

Grade was also significant, $F(1, 41) = 14.29$, $MSE = 0.173$, $p < .001$. Older fifth grade children identified objects at a more fragmented level (better performance) than younger second grade children (5.51 vs. 5.98, respectively).

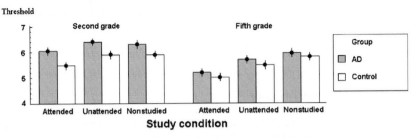

Figure 3. Mean completion level (mean correct identification threshold) necessary for picture identification as a function of group, grade, and study condition (attended, unattended, and nonstudied). Bars indicate the standard error of the mean (*SEM*).

Threshold

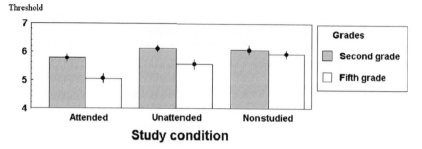

Figure 4.   Study condition × Grade interaction in the implicit memory test collapsed across group. Bars indicate the standard error of the mean (*SEM*).

Figure 4 shows the interaction between study condition and grade, $F(2, 82) = 5.76$, $MSE = 0.116$, $p < .005$. Simple effect contrasts among the means showed that attended stimuli were identified at a more fragmented level (a lower threshold) than nonstudied stimuli for second and fifth grade children, $F(1, 22) = 11.3$, $p < .001$. However, only fifth grade children showed a significant difference for unattended pictures compared to nonstudied ones, $F(1, 22) = 3.93$, $p = .03$. The identification threshold for unattended and nonstudied pictures did not differ in second grade children, $F(1, 22) = 0.59$, $p = .38$. No other two-way interaction was statistically significant. The two-way interaction of attention condition and group was not significant ($p > .05$). The difference between attended and nonstudied pictures for AD children collapsed across grade was highly significant, $F(1, 20) = 14.529$, $MSE = 0.140$, $p < .001$. AD children showed perceptual priming for attended pictures at study. The three-way interaction Attention condition × Group × Grade was also not significant.

We conducted an ANOVA on the percentage of errors corresponding to the study phase with group and grade as variables. The total mean of errors was 11.7%. Group (AD and control children) was not significantly different ($p = .24$); grade was significant ($p < .000$), younger children had more errors than older children (16.6% of errors vs. 6.8%, respectively). The interaction Group × Grade was not significant ($p > .05$). These results suggest that the developmental effect was present during the study phase as well.

Explicit recognition test

The results of the recognition test were expressed as the sensitivity measure of the Signal Detection Theory (d') defined in terms of the distances between the Noise (to say "yes" to a new drawing) and the Signal + Noise distributions (to say "yes" to an old drawing). The results of the recognition

TABLE 1

Explicit memory performance in d' and c values for attended and unattended pictures of AD and control children in second and fifth grades

| Grade | Attended | | Unattended | |
	AD	Control	AD	Control
Second	0.892 (0.00)	0.930 (−0.126)	0.104 (0.434)	0.369 (0.155)
Fifth	1.500 (0.209)	1.826 (0.286)	0.343 (0.809)	0.477 (0.961)

AD = attention deficit children; d' = Sensitivity; c = criterion (in brackets).

task (d' and c parameters of the Signal Detection Theory) as a function of study condition, group, and grade are shown in Table 1.

To calculate the d' values, we used the SDT_SP program (Reales & Ballesteros, 1994). The ANOVA performed on d' scores for two study conditions (attended vs. unattended) × 2 grades (second vs. fifth) × 2 groups of children (AD vs. control) showed a significant main effect of study condition, $F(1, 41) = 92.62$, $MSE = 0.224$, $p < .001$. As expected, recognition performance for attended pictures was better than for unattended pictures ($d' = 1.287$ and 0.824, respectively). The main effect of grade was also significant, $F(1, 41) = 4.452$, $MSe = 1.076$, $p < .04$. Explicit recognition was higher for the older ($d' = 1.037$) than for the younger children ($d' = 0.574$). This result is not surprising as explicit memory improves with age until adulthood and then diminishes during ageing. Group was not statistically significant ($d' = 0.700$ and $d' = 0.910$, for AD children and normal children, respectively; $p > .05$). However, the interaction between study condition and grade was highly significant, $F(1, 41) = 8.328$, $MSE = 0.224$, $p < .006$. The simple effects analysis of the interaction indicated that both second and fifth grade children remember attended pictures better ($d' = 0.911$ and $d' = 1.663$, respectively; $p < .009$) than unattended pictures ($d' = 0.237$ and $d' = 0.410$, respectively; $p = .47$) but the advantage for attended pictures was more pronounced for the older than for the younger children. No other interaction was significant. The results are interesting because they show the improvement of attention and episodic memory with age. As children grow older, their performance in the explicit recognition task for pictures that were selectively attended during the encoding episode improved significantly.

DISCUSSION

The present study yielded the following main findings. First, AD children as well as control children at both grades showed perceptual priming for pictures encoded under selective attention conditions. Second, the picture identification threshold was higher for AD children than for control children; that is, AD children needed more perceptual information available on the screen before being able to identify the stimulus. Third, the older groups (AD and normal children) identified the pictures at a lower threshold (a more fragmented level) than the younger groups showing better performance. Fourth, both grades showed perceptual priming for attended pictures but only fifth grade children (AD and control children) showed some facilitation for unattended pictures. Fifth, pictures attended at encoding were remembered more accurately than unattended pictures in all groups but as expected, performance on the recognition task improved with age. Finally, the effect of attention on recognition increased with age as the advantage for attended pictures was superior for the older than for the second grade children.

The finding that AD and non-AD children at both grades showed significant perceptual priming in the implicit PFCT is consistent with the results from two recent studies (Aloise et al., 2004; Burden & Mitchell, 2005). These results suggest that perceptual implicit memory appears to be normal in AD children. In addition, the present study is the first to report that perceptual implicit memory for pictures in second and fifth grade AD and control children is affected by attention at encoding. These results are consistent with recent findings with young adults (Ballesteros, Reales, García, & Carrasco, 2006) and older healthy adults (Ballesteros, Reales, & Mayas, 2006). Implicit memory as well as explicit recognition tasks are affected similarly by attention at encoding, indicating that implicit memory is not automatic and requires attention. Several implicit memory studies conducted with adult participants using words as stimuli have concluded also that attention is necessary at study to form a representation that can support perceptual priming (e.g., Crabb & Dark, 1999; McDonald & MacLeod, 1998). Moreover, these studies did not report the dissociation between implicit and explicit memory tasks.

The present study showed a substantial improvement from 7–8 years to 10–11 years in children's ability to integrate spatially separated information into complete pictures as shown by their performance in the PFCT. This improvement in performance with age was shown in both AD children and control children as both groups showed a reliable developmental trend. The present results are congruent with findings from a previous study conducted with normal participants from 5 to 28 years. In the study, Cycowicz et al. (2000) found also that the identification threshold increased as age decreased; that is, the younger the child, the more information was needed to identify the fragmented picture.

Other previous studies using hierarchical letter and geometric forms and patterns have shown that even 4-year-olds are able to integrate the parts of spatial arrays to form a coherent whole and that this ability improves with age. Six-year-olds' performance was better and less easily disrupted than that of the younger children and this integration process undergoes developmental changes up to the age of 8 (Dukette & Stiles, 1996, 2001). A recent study has shown that organisation processes did not develop at the same rate. Visual search and speeded classification tasks' performance with few-element patterns, but not with many-element patterns, showed a developmental trend. On the other hand, performance for local targets improved with age for many-element patterns but not for few-element patterns (Kimchi et al., 2005). Our results in the PFCT showed that AD children perform more poorly than control children of the same age. The main effect of diagnostic condition on fragmentation thresholds made us speculate that the slower developmental trajectory of AD children compared to their control peers is due to the attentional demands of integrating different bits of information into meaningful and coherent pictures.

As far as we know, this is the first study that has assessed the influence of selective attention at encoding in perceptual priming and explicit recognition tests in AD children and normal control children at two different ages. Children in our study could select and process effectively one of the two superimposed outline drawings using colour as an effective selective cue. The effect of attention at encoding was similar in AD and control children. The preservation of perceptual implicit memory for attended pictures in young and older AD children is important as it shows that these children exhibited similar levels of perceptual facilitation as control did. However, it is important to note that AD children needed more information available in the visual display (higher threshold) than control children did before being able to identify the picture. This happens not only for attended pictures at study but for unattended and nonstudied pictures as well. In other words, AD children exhibited a general deficit instead of a specific deficit in the PFCT. Interestingly, the difference between normal and AD groups was reduced with age. Eleven-year-old AD children not only improved general performance in the perceptual task but also reduced their difference compared with normal control children.

The finding of reduced priming for unattended pictures relative to attended ones shown by the older children in this study is consistent with previous results from our laboratory using the same paradigm with young adult participants assessed at different delay conditions after encoding (Ballesteros, Reales, García, & Carrasco, 2006). At all delay conditions (immediate, 5 min, 1 day, 1 week, and 1 month), picture priming identification threshold was lower for the attended than for the unattended pictures, and attended pictures were recognised more accurately than unattended pictures. The findings from the present study are also congruent with those from another recent study conducted with young, older healthy adults, and Alzheimer's disease patients. This study showed significant facilitation for attended compared to nonstudied pictures in young and older healthy adults (Ballesteros, Reales, & Mayas, 2006). So far, these studies conducted with AD children, normal children, young adults, and older healthy adults indicate that perceptual implicit memory as well as recognition require attention at the time of encoding. Implicit memory assessed by the PFCT was similar in both groups. Moreover, AD children and control children did not differ in recognition as both performed similarly on the explicit memory test.

Attention at encoding improves perceptual priming compared to unattended stimuli. Our results for attended stimuli are in agreement with those of Burden and Mitchell (2005; Burden, 2002). In the Burden and Mitchell study, the performance of children with and without attentional problems was examined under full attention conditions. These authors found that AD and control children did not differ when implicit memory was assessed with a perceptual memory test or with a recognition test. However, these researchers obtained group differences on a conceptual implicit memory test (category exemplar generation with words and pictures). Our results in recognition also indicated that AD and control children performed similarly. Age-related improvements occurred in AD and non-AD children. In the present study, using an overlapping picture encoding task to manipulate selective attention we have further shown that attention at encoding matters not only for explicit memory tests but also for implicit perceptual memory tests that have been considered automatic at least to a certain degree (Parkin & Russo, 1990; Szymanski & McLeod, 1996).

The role of extrastriate areas in perceptual priming

Cognitive neuroscience has been very useful to study the memory systems as well as the attention networks in the brain. At a clinical level, a recent fMRI study on brain activation differences between 12 control and 12 AD children (9- to 12-year-olds) has shown limited group differences in the performance of a selective attention task (Booth et al., 2005). Both control and AD children showed activation in the regions of interest (the superior parietal lobe and lateral premotor network) for a visual selective attention task. In contrast, the results showed a widespread hypoactivity in the clinical group during response inhibition measured by a go/not go task in the frontostriatal network. This hypoactivity was not found in the control group. Decreased blood flow and energy utilisation occurred in the prefrontal cortex and striatum of AD children (Booth et al., 2005; Paule et al., 2000). The deficits reported in these children are similar to those shown by adults suffering from frontal lobe damage and implicate executive function deficits. These findings suggest that the frontal cortex is not working adequately in AD children (Faraone & Biederman, 1998). In contrast, more posterior cortical areas such as the occipital cortex may work properly in these children. This may explain why AD children show normal perceptual priming for pictures.

As suggested by behavioural and neuroimaging studies, areas in the occipital cortex are involved not only in visual but also in haptic object priming (Ballesteros & Reales, 2004; James et al., 2002; James, Harman James, Humphrey, & Goodale, 2006; Sathian, 2005). Moreover, according to Amedi, Kriegstein, van Atteveldt, Beauchamp, and Naumer (2005), cortical areas that were once believed to be modality specific can be activated when stimuli are presented to visual, auditory, or tactual modalities.

Imaging studies with AD children suggested a decreased blood flow in the prefrontal cortex and striatum (Paule et al., 2000). These abnormalities have not been reported in more posterior cortical areas such as those occipital areas involved in visual (and haptic) object priming. This may explain why children with AD disorders exhibited normal perceptual priming for visually attended objects at encoding in the present study. Recently, Burden and Mitchell (2005) have reported a memory deficit in AD children compared to control children of the same age in a conceptual implicit memory task that is likely mediated by the frontal lobes. This result can be readily explained by some imaging studies that suggested a frontal systems dysfunction in AD children (e.g., Booth et al., 2005; Paule et al., 2000). The involvement of the frontal and prefrontal cortical areas on conceptual priming tasks as well as the dependency of perceptual priming tasks on occipital areas of the brain was suggested by Cabeza and Nyberg (2000).

CONCLUSION

The present study demonstrated significant perceptual priming effects for attended objects at encoding for both age groups and condition (AD and control schoolchildren); that is, picture-fragment identification thresholds were lower for the attended than for new pictures, independently of group and school grade. However, perceptual priming was not found for unattended pictures in second grade children. Only fifth grade children showed reliable but reduced perceptual priming effects for unattended pictures in comparison to attended pictures as has been found in a series of previous experiments conducted with adult participants (Ballesteros, Reales, García, & Carrasco, 2006). The results support the idea that implicit memory is not automatic and requires attention at encoding. AD children as well as control children were able to use selective attention to more thoroughly encode one picture over the other and this additional encoding produced better performance (a lower threshold) on a PFCT used to measure perceptual implicit memory. Not surprisingly, the attended pictures were also better recognised than the unattended pictures by young and older schoolchildren. Children at both grades remember attended pictures more accurately but the advantage for attended pictures was more pronounced for the older than for the younger children.

The findings showed that there was neither a perceptual implicit nor an explicit recognition memory deficit in AD children. However, AD children performed overall more poorly than control children in the implicit task. To be able to identify the pictures, younger second grade AD children required considerably more picture information in the visual field than control children of the same age. The difference, however, decreased by the age of 11, suggesting that substantial developmental changes have occurred in children's ability to integrate spatially separated visual elements into complete pictures of familiar objects. The implicit memory for attended stimuli (as a measure of implicit memory) observed in AD children may have important implications. This preserved memory ability may be used to try to teach these children other complex cognitive skills using, for example the method of the "vanishing cues". This method has been successfully applied to train an amnesic patient with preserved implicit memory to perform rather complex tasks (see Glisky, Schacter, & Tulving, 1986; Schacter, 1996, for details).

REFERENCES

Achenbach, T. M., & Edelbrock, C. (1983). *Manual for the Child Behavior Checklist and Revised Child Behavior Profile.* Burlington, VT: University of Vermont.

Aloise, B. A., McKone, E., & Heubeck, B. G. (2004). Implicit and explicit memory performance with attention deficit/hyperactivity disorder. *British Journal of Developmental Psychology, 22,* 275–292.

Amedi, A., Kriegstein, K., van Atteveldt, N. M., Beauchamp, M. S., & Naumer, M. J. (2005). Functional imaging of human crossmodal identification and object recognition. *Experimental Brain Research, 166,* 559–571.

American Psychiatric Association. (1994). *Diagnostic and statistical manual of mental disorders* (4th ed.). Washington, DC: Author.

Ballesteros, S., González, M., Mayas, J., Reales, J. M., García, B., & de Paz, S. (in press). Crossmodal object priming in young and older adults: Multisensory processing in vision, touch, and audition. *Manuscript submitted for publication.*

Ballesteros, S., & Reales, J. M. (1998). Influencia de la atención en la memoria implícita y explícita [The influence of attention in implicit and explicit memory]. In J. Botella & V. Ponsoda (Eds.), *La atención: Un enfoque pluridisciplinar* (pp. 237–250). Valencia, Spain: Promolibro.

Ballesteros, S., & Reales, J. M. (2004). Haptic priming in normal aging and Alzheimer's disease: Evidence for dissociable memory systems. *Neuropsychologia, 42,* 1063–1070.

Ballesteros, S., & Reales, J. M. (2005). Influencia de la atención selectiva en la memoria implícita de objetos. In J. J. Ortells, C. Noguera, E. Carmona & M. T. Daza (Eds.), *La atención: Un enfoque multidisciplinar [Attention: A multidisciplinary approach]* (Vol. III, pp. 135–147). Valencia, Spain: Promolibro.

Ballesteros, S., Reales, J. M., García, E., & Carrasco, M. (2006). Selective attention affects implicit and explicit memory for familiar objects at different delay conditions. *Psicothema, 18,* 96–107.

Ballesteros, S., Reales, J. M., & Manga, D. (1999). Implicit and explicit memory for familiar and novel objects presented to touch. *Psicothema, 11*, 785–800.

Ballesteros, S., Reales, J. M., & Mayas, J. (2006). Implicit memory and selective attention at encoding: Effects on aging and dementia. In S. Ballesteros (Ed.), *Aging, cognition, and neuroscience* (pp. 39–41). Madrid, Spain: UNED Varia.

Ballesteros, S., Reales, J. M., & Mayas, J. (2007). Picture priming in aging and dementia. *Psicothema, 19*, 239–244.

Barkley, R. A. (1997). Behavioral inhibition, sustained attention, and executive functions: Constructing a unifying theory of ADHD. *Psychological Review, 121*, 65–94.

Barkley, R. A. (1998). *Attention-deficit hyperactivity disorder: A handbook for diagnosis and treatment* (2nd edn). New York: Guilford Press.

Biederman, J. (1998). Attention-deficit/hyperactivity disorder: A lifespan perspective. *Journal of Clinical Psychiatry, 59*(Suppl. 7), 4–16.

Biederman, I., & Cooper, E. E. (1992). Size invariance in visual object priming. *Journal of Experimental Psychology: Human Perception and Performance, 18*, 121–133.

Booth, J. R., Burman, D. D., Meyer, J. R., Lei, Z., Trommer, B. L., Davenport, N. D., et al. (2005). Larger deficits in brain networks for response inhibition than for visual selective attention in attention deficit hyperactivity disorder (ADHD). *Journal of Child Psychology and Psychiatry, 46*, 94–111.

Burden, M. J. (2002). Implicit and explicit memory in children with and without attention deficit hyperactivity disorder. *Dissertation Abstract International, Section B: The Sciences and Engineering, 63*(6B), 3039.

Burden, M. J., & Mitchell, D. B. (2005). Implicit memory development in school-aged children with attention deficit hyperactivity disorder (ADHD): Conceptual priming deficit? *Developmental Neuropsychology, 28*, 779–807.

Cabeza, R., & Nyberg, L. (2000). Imaging cognition II: An empirical review of 275 PET and fMRI studies. *Journal of Cognitive Neuroscience, 12*, 1–47.

Cooper, L. A., Schacter, D. L., Ballesteros, S., & Moore, C. (1992). Priming and recognition of transformed three-dimensional objects: Effects of size and reflection. *Journal of Experimental Psychology: Learning, Memory, and Cognition, 18*, 43–57.

Cornoldi, C., Barbieri, A., Gaiani, C., & Zocchi, C. (1999). Strategic memory deficits in attention deficit disorder with hyperactivity participants: The role of executive processes. *Developmental Neuropsychology, 15*, 53–71.

Crabb, B. T., & Dark, V. J. (1999). Perceptual implicit memory requires attentional encoding. *Memory and Cognition, 27*, 267–275.

Craik, F. I. M., Govoni, R., Naveh-Benjamin, M., & Anderson, N. D. (1996). The effects of divided attention on encoding and retrieval processes in human memory. *Journal of Experimental Psychology: General, 125*, 159–180.

Cycowicz, Y. M., Friedman, D., Snodgrass, J. G., & Rothstein, M. (2000). A development trajectory in implicit memory is revealed by picture fragment completion. *Memory, 8*, 19–35.

DiGiulio, D. V., Seidenberg, M., O'Leary, D. S., & Raz, N. (1994). Procedural and declarative memory: A developmental study. *Brain and Cognition, 25*, 79–91.

Drechsler, R., Brandeis, D., Földényi, M., Imhof, K., & Steimhausen, H.-C. (2005). The course of neuropsychological functions in children with attention deficit hyperactivity disorder from late childhood to early adolescence. *Journal of Child Psychology and Psychiatry, 46*, 824–836.

Dukette, D., & Stiles, J. (1996). Children's analysis of hierarchical patterns: Evidence from a similarity judgment task. *Journal of Experimental Child Psychology, 63*, 103–140.

Dukette, D., & Stiles, J. (2001). The effects of stimulus density on children's analysis of hierarchical patterns. *Developmental Science, 4*, 233–251.

Faraone, S. V., & Biederman, J. (1998). Neurobiology of attention-deficit hyperactivity disorder. *Biological Psychiatry, 44*, 951–958.

Fisk, A. D., & Schneider, W. (1984). Memory as a function of attention, level of processing, and automatization. *Journal of Experimental Psychology: Learning, Memory, and Cognition, 10,* 181–196.

Fusari, A., & Ballesteros, S. (2006, June). *Olfactory perceptual priming is resistant to aging and long-lasting.* Poster presented at the seventh annual meeting of the International Multisensory Forum, Dublin, Ireland.

Glisky, E. L., Schacter, D. L., & Tulving, E. (1986). Learning and retention of computer- related vocabulary in memory impaired patients: Method of vanishing cues. *Journal of Clinical and Experimental Neuropsychology, 3,* 292–312.

Goldstein, E. B., & Fink, S. I. (1981). Selective attention in vision: Recognition memory for superimposed line drawings. *Journal of Experimental Psychology: Human Perception and Performance, 7,* 954–967.

James, T. W., Harman James, K., Humphrey, G. K., & Goodale, M. A. (2006). Do visual and tactile object representations share the same neural substrate? In M. A. Heller & S. Ballesteros (Eds.), *Touch and blindness: Psychology and neuroscience* (pp. 139–156). Hillsdale, NJ: Lawrence Erlbaum Associates, Inc.

James, T. W., Humphrey, G. K., Gati, J. S., Servos, P., Menon, R. S., & Goodale, M. A. (2002). Haptic study of three-dimensional objects activates extrastriate visual areas. *Neuropsychologia, 40,* 1706–1714.

Kimchi, R., Hadad, B., Behrmann, M., & Palmer, S. E. (2005). Microgenesis and ontogenesis of perceptual organization: Evidence from global and local processing of hierarchical patterns. *Psychological Science, 15,* 282–290.

Lorsbach, T. C., & Worman, L. J. (1989). The development of explicit and implicit forms of memory in leaning disabled children. *Contemporary Educational Psychology, 14,* 67–76.

MacDonald, P. A., & MacLeod, C. M. (1998). The influence of attention at encoding on direct and indirect remembering. *Acta Psychologica, 98,* 291–310.

Manga, D., Garrido, I., & Pérez-Solís, M. (1997). *Atención y motivación en el aula [Attention and motivation in the classroom].* Madrid, Spain: Europsyque.

McDonald, S., Bennett, K. M. B., Chambers, H., & Castiello, U. (1999). Covert orienting and focusing of attention in children with attention deficit hyperactivity disorder. *Neuropsychologia, 37,* 345–356.

Mitchell, D. B. (1989). How many memory systems? Evidence from aging. *Journal of Experimental Psychology: Learning, Memory, and Cognition, 15,* 31–49.

Mitchell, W. G., Chavez, J. M., Baker, S. A., Guzman, B. L., & Azen, S. P. (1990). Reaction time, impulsivity, and attention in hyperactive children a control: A video game technique. *Journal of Child Neurology, 5,* 195–204.

Mulligan, N. W. (1998). The role of attention during coding on implicit and explicit memory. *Journal of Experimental Psychology: Learning, Memory, and Cognition, 24,* 27–47.

Park, S. M., Gabrieli, J. D., Reminger, S. L., Monti, L. A., Fleischman, D. A., Wilson, R. S., et al. (1998). Preserved priming across study–test picture transformations in patients with Alzheimer's disease. *Neuropsychology, 12,* 340–352.

Parkin, A. J., & Russo, R. (1990). Implicit and explicit memory and the automatic/effortful distinction. *European Journal of Cognitive Psychology, 2,* 71–80.

Parkin, A. J., & Streete, S. (1988). Implicit and explicit memory in young children and adults. *British Journal of Psychology, 79,* 361–369.

Paule, M. G., Rowland, A. S., Ferguson, S. A., Chelonis, J. J., Tannock, R., Swanson, J. M., & Castellanos, F.X. (2000). Attention deficit/hyperactivity disorder: Characteristics, interventions, and models. *Neurotoxicology and Teratology, 22,* 631–651.

Posner, M. I. (1980). Orienting attention. *Quarterly Journal of Experimental Psychology, 32,* 3–25.

Reales, J. M., & Ballesteros, S. (1994). SDT_SP, a program in Pascal for computing parameters and significance tests from several detection theory designs. *Behavioral Research Methods, Instruments, and Computers, 26,* 151–155.

Reales, J. M., & Ballesteros, S. (1999). Implicit and explicit memory for visual and haptic objects: Cross-modal priming depends on structural descriptions. *Journal of Experimental Psychology: Learning, Memory, and Cognition, 118,* 219–235.

Rock, I., & Gutman, D. (1981). The effect of inattention on form perception. *Journal of Experimental Psychology: Human Perception and Performance, 7,* 275–287.

Russo, R., Nichelly, P., Gibertoni, M., & Cornia, C. (1995). Developmental trends in implicit and explicit memory: A picture completion study. *Journal of Experimental Child Psychology, 59,* 566–578.

Sathian, K. (2005). Visual cortical activity during tactile perception in the sighted and the visually deprived. *Developmental Psychobiology, 46,* 279–286.

Schacter, D. L. (1987). Implicit memory: History and current status. *Journal of Experimental Psychology: Human Learning and Memory, 13,* 501–518.

Schacter, D. L. (1996). *Searching for memory: The brain, the mind, and the past.* New York: Basic Books.

Schacter, D. L., Cooper, L. A., & Delaney, S. M. (1990). Implicit memory for unfamiliar objects depends on access to structural description. *Journal of Experimental Psychology: General, 119,* 5–24.

Snodgrass, J. G., Smith, B., Feenan, K., & Corkin, J. (1987). Fragmenting pictures on the Apple Macintosh computer for experimental and clinical applications. *Behavior Research, Methods, Instruments and Computers, 19,* 270–274.

Snodgrass, J. G., & Vanderwart, M. (1980). A standardized set of 260 pictures: Norms for name agreement, image agreement, familiarity, and visual complexity. *Journal of Experimental Psychology: Human Learning and Memory, 6,* 807–815.

Snyder, J., Prichard, J., Schrepferman, L., Patrick, M. R., & Stoolmiller, M. (2004). Child impulsiveness-inattention, early peer experiences, and the development of early onset conduct problems. *Journal of Abnormal Child Psychology, 32,* 579–594.

Sonuga-Barke, E. J. S., de Houwer, J., de Ruiter, K., Ajzenstzen, M., & Holland, S. (2004). AD/HD and the capture of attention by briefly exposed delay-related cues: Evidence from a conditioning paradigm. *Journal of Child Psychology and Psychiatry, 45,* 274–283.

Szymanski, K. F., & MacLeod, C. M. (1996). Manipulation of attention at study affects an explicit but not an implicit test of memory. *Consciousness and Cognition, 5,* 165–175.

Thurstone, L. L., & Yela, M. (1968). *Faces Differences Perception Test.* Madrid, Spain: TEA.

Weiss, G., & Hechtman, L. (1993). *Hyperactive children grown up* (2nd ed). New York: Guilford Press.

EUROPEAN JOURNAL OF COGNITIVE PSYCHOLOGY
2007, 19 (4/5), 628–649

Strategic knowledge and consistency in students with good and poor study skills

Chiara Meneghetti, Rossana De Beni, and Cesare Cornoldi

Department of General Psychology, University of Padova, Padova, Italy

Early adolescence is characterised by an increase in study requirements and the establishment of a systematic study method. However some students fail in study tasks. Teachers often attribute their difficulties to poor content knowledge or poor effort, without taking into consideration the specific role of study strategies. The present paper tests the hypothesis that poor study skills are related to students' inadequate knowledge of good strategies and/or to their inconsistent use. From a sample of 354 students, aged between 12 and 15, on the basis of a study standardised test (AMOS 8–15; Cornoldi, De Beni, Zamperlin, & Meneghetti, 2005) we selected two groups of students, with good and poor study skills respectively, and we asked them to rate their knowledge and actual use of 22 good and 10 less adequate study strategies. We found that all students reported using strategies to a lesser extent than should be expected on the basis of their estimated importance, but they were all able to distinguish between poor and good strategies. However, students with poor study skills were less able to make this distinction and were less consistent in matching their knowledge to their use of strategies. It is concluded that strategic use and consistency play a crucial role in successful studying.

During adolescence, students typically meet new study requirements, which will progressively increase when entering university. However, some students fail on study performance. Usually teachers attribute school failures to the fact that students are not prepared and do not study enough.

However, a growing number of studies have analysed to what degree cognitive and metacognitive aspects, and their relationship, influence school achievement. Cognitive processing includes skills that help learners carry out a specific study task (such as attention, language, memory, etc.); metacognitive aspects include skills that help learners understand and regulate these cognitive processes (Artzt & Armour-Thomas, 1998). There exists a large

Correspondence should be addressed to Cesare Cornoldi, Dipartimento di Psicologia Generale, Via Venezia, 8, 35131 Padova, Italy. E-mail: cesare.cornoldi@unipd.it

The authors wish to thank Nicolette Whitteridge for helpful comments on this paper.

body of evidence suggesting that metacognitive processing is a hallmark of effective learning. Pressley and his colleagues (van Etten, Pressley, & Freebern, 1998; van Meter, Yokoi, & Pressley, 1994), for example, concluded that successful learners are motivationally well oriented and actively use metacognitive strategy knowledge to manage their coursework.

In the realm of metacognitive studies there is an increasing interest in analysing the role of study learning strategies with particular reference to the case of study processes, a case especially representative for examining the relationship between strategies and achievement. Study strategies are defined as a group of systematic procedures or activities applied during learning that support students' active manipulation of text content and other material (figures, tables, etc.). In fact, studying is characterised by a combination of study strategies used in different phases, such as planning, reading comprehension, memorisation, and review phases (Pressley et al., 1995; Schneider & Pressley, 1997).

Several researches have focused on the development of study strategies during childhood and adolescence. In particular researches comparing study methods and strategies among adolescents found that high school achievers used better adaptive strategies with favourable consequences on school achievement (Wolters, 1998). For example in reading comprehension high achievers showed greater organisational abilities (Kleijn, van der Ploeg, & Topman, 1994), a greater tendency to use previous knowledge to understand the text (Staynoff, 1997); moreover they were better able to distinguish the main ideas from the details (Moreland, Dansereau, & Chmielewski, 1997). High achievers also actively memorised the content (Beishiuzen & Stoutjesdijk, 1999), using schema-driven strategies involving schemas, diagrams, tables, and note making (Wood, Motz, & Willoughby, 1998). During the review phase they paid attention to revision of the content and used self-testing strategies to verify their learning (Wilding & Valentine, 1992).

Some studies found that study strategies are strictly related to the self-regulation ability (Cornoldi, De Beni, & Fioritto, 2003; Moè, Cornoldi, De Beni, & Veronese, 2004), i.e., the metacognitive control of the study activity. Cornoldi et al. (2003) showed that self-regulation is a relevant factor in successful studying. Using a structural equation modelling methodology, the authors found that strategies and self-efficacy (Bandura, 1986) abilities modulated the self-regulation factor, which together with motivational aspects (achievement goals and effort attribution) were good predictors of academic achievement. Similarly, Bembenutty and Zimmerman (2003) found that motivational aspects (i.e., self-efficacy, intrinsic interest, and outcome expectancy) influenced the final course grade, through the mediation of self-regulation. According to these authors, self-regulation represents the metacognitive ability regulating the use of study strategies. At the same time other studies showed the relevance of knowledge about efficacy of strategies. Some of them (e.g., Hofer & Pintrich, 2004; Schraw, 1998) examined the role of conditional knowledge, i.e., of metacognitive knowledge focusing of the subject's perception of the specific utility of a strategy for him-/herself.

Some metacognitive studies focused on the distinction between reported "knowledge about the utility" and "use" of study strategies. The "knowledge about the utility" of study strategies involves the ability to recognise adequate and inadequate strategies that can be applied in the different learning phases. Some authors (Garner, 1990; Nolen, 1988; Nolen & Flaladyna, 1990) found that in order to have a good study method, it is important to evaluate also the personal utility of study strategies. Utility refers to the students' personal and informal assessment of the usefulness of a particular learning strategies or method for their own study. Student knowledge about utility, on the other hand, should describe metacognitive beliefs about studying and refers to the ideal perception of self as student (Higgins, 1987; Markus & Nurius, 1986). Therefore, utility ratings could be considered as an index of a particular subsystem of the students' "ideal" self (describing how the student would like to study) or rather of the "imperative" self (describing how the student believes he/she should study). If students do not find ways to internalise a particular learning strategy and to apply it in study activities, they will not use it. The "use" of studying strategies is the ability to utilise them monitoring and regulating their application in the studying phases (Cornoldi, 1995; De Beni & Moè, 1997; De Beni, Moè, & Cornoldi, 2003; Moè, Cornoldi, & De Beni, 2001). The reported use should describe the actual perception of self as students, i.e., the actual behaviour in study activity. A characteristic of successful students is their ability to model their awareness on the utility and flexibility of strategies during study. They are aware of the different study strategies, choose the most appropriate ones, and monitor their use during learning. The latter aspect is generally considered a critical factor for study skills abilities, because it provides learners with feedback regarding their progress in performance; without self-monitoring, efficient control over one's cognitive system may be very limited (Butler & Winne, 1995; Pintrich, Wolters, & Baxter, 2000).

A series of studies have already shown that effective use of strategies, associated with knowledge about them, affects study performance (e.g., Pintrich & Schunk, 1996; Schunk & Zimmerman, 1998; Weinstein & Mayer, 1986). For example, Ruban, McCoach, McGuire, and Reis (2003) investigated whether the perceived usefulness and the use of self-regulated learning strategies provided a differential prediction on academic achievement in students with and without learning disabilities (LD). Using a structural equation modelling, they found that students with LD did not differ significantly from students without LD in the perception of usefulness but differed in the use of self-regulated strategies. For students with LD, academic achievement seemed to be mediated by compensatory strategies due to technology support and help from other people, but not by the use of self-regulated strategies. If it is true that poor achievement is in general associated with poor knowledge and use of strategies, it is also true that not all the procedures/strategies implemented by students are equally effective. For example in stark contrast with good achievers, low achievers tend to use maladaptive strategies characterised by strict adherence to the text and reduced personal elaboration of the content (Gadzella, 1995). They have difficulty weighting the importance of the different information in a text, and adopt underlining or read-and-repeat techniques (Wood et al., 1998).

Not only does knowledge about the utility and use of strategies seem important, but there is also a third aspect, i.e., strategic consistency, which expresses the correspondence between knowledge about the utility and use ratings of the same strategy (De Beni & Moè, 1997; Moè et al., 2001). The notion of strategic consistency can be considered as an example of the case when a person sees his/her real self not far from his/her ideal self (Higgins, 1987; Markus & Nurius, 1986) and can be applied not only to students who give both high ratings for knowledge about the utility and for use of the same strategy but also to students who give low ratings both in the case of knowledge and of use. A smaller distance between these two elements reflects good strategic consistency, while a greater distance indicates considerable strategic inconsistency between the knowledge and the use of a strategy. Accordingly Moè et al. (2001) analysed the difference between knowledge, use ratings, and the correspondence between these ratings (strategic consistency index) in low and high university achievers. Academic achievement was measured using the number of the exams passed. Results clearly showed that low achievers present higher discrepancy between knowledge and use ratings, i.e., a greater inconsistency in comparison with high achievers. Taken together these results suggest that strategic aspects, articulated in knowledge, use, and strategic consistency, could be critical metacognitive factors in students' success.

Although several studies investigated the relation between knowledge and use of strategies (e.g., Pintrich & Schunk, 1996; Pressley et al., 1995), no study directly used a strategic consistency index to shed light on study success in early adolescents. Given the relevant role of study strategies consistency in the success of university students (Moè et al., 2001), the general goal of the present study was to explore whether this metacognitive aspect could be crucial in adolescent students as well.

Early adolescence represents a critical moment for the establishment of a good study method, which should then be used when confronted with the more complex study requirements that students will meet in the following years. Furthermore, the changes in the classroom environment during transition from primary to secondary school affect children's perceptions of their ability to master study material (e.g., Eccles, Roeser, Wigfield, & Freedman-Doan, 1999). Despite teachers' typical assumption that study success is mainly due to intelligence, knowledge, and effort, strategic abilities, and in particular strategic consistency, could be a critical factor that differentiates students with good and poor study skills.

In conclusion, despite the general idea that less successful students are poor strategy users, so far no studies have specifically examined either low success students, i.e., students who fail in study/learning tasks, or the nature of strategic deficit in adolescents with poor study skills. The aim of the present study was to compare adolescent students with good and poor study skills in (1) knowledge of efficacy of strategies, (2) their reported use, and (3) the strategic consistency in the use of good and less effective strategies.

In the present study we also considered the distinction between good strategies (e.g., using schemas, writing notes beside the text) and inadequate strategies (e.g., reading the text aloud or skipping difficult content) in knowledge ratings, use ratings, and consistency scores. We supposed that students with high study skills would have developed greater ability to distinguish between good and less effective strategies as compared with low study skills students, and we attributed greater importance to good strategies both for utility and use ratings. By contrast, we expected that students with low study skills would not only have more difficulty in distinguishing between good strategies and less effective strategies in their utility, but would also consider as less important the use of effective study strategies. In accordance with these hypotheses we expected that low study skills would be also less consistent than high study skills students in the use of good and less effective strategies. As suggested by Moè et al. (2001), students with low study skills should have a more confused representation of the utility of strategies and consequently should be less systematic in their reported use of strategies independently of their knowledge of the efficacy of strategies. In fact, we predicted that students with low study skills would present a tendency to lower consistency in the reported use of the strategies, applied, to a greater or lesser extent, independently from the perceived efficacy.

These hypotheses were tested using an objective Study Task (ST) and two Strategy Questionnaires (SQ1 and SQ2) included in a standardised battery, recently devised in Italy, which measures the components of study abilities (Cornoldi, De Beni, Zamperlin, & Meneghetti, 2005). Study Task (ST) measures the ability to learn text content. Time is given to study a text after which recall is tested using three tasks: The first measures the ability to select relevant information, the second and third measure the recall of information recognising true/false statements (second task) and giving brief open answers (third task). Tasks are chosen which require both multiple choice and open answers to check for a possible negative effect of multiple choice test (March, Roediger, Bjork, & Bjork, 2005).

The Strategy Questionnaires measure the knowledge (SQ1) and the use (SQ2) of adequate ("good") and inadequate ("less good") strategies and makes it possible to calculate the correspondence between these two ratings (strategic consistency; Moè et al., 2001). Questionnaires include 22 "good" strategies and 10 "less good" strategies. The assumption underlying the strategy consistency score is that effective students should use all of the strategies they consider useful, in a flexible way depending on the particular task and content characteristics.

METHOD

Participants

354 students aged between 12 and 15 (239 males and 115 females) participated in the initial screening necessary for the selection of groups with good and poor study skills. Students attended different types of schools and were representative of student population of this age in North-Eastern Italy. Gender and age information concerning the participants, divided into two study skill groups, are presented in Table 1 (see Student Selection section).

Materials

Strategy questionnaires.   The two AMOS 8–15 Strategy Questionnaires (Cornoldi et al., 2005) collect ratings of the knowledge (SQ1) and of the use (SQ2) of 32 study strategies (see Table 4 for the complete list of studying strategies). The strategies are listed in a different order in the two questionnaires. The instructions for SQ1 and SQ2 are different (see Procedure section). The ratings are given using a Likert scale from 1 (no knowledge/no use) to 4 (good knowledge/good use). The internal consistency (Cronbach's alpha) calculated on current sample is .68 for the knowledge rating, .74 for the use rating, and .71 for the strategic consistency score, that are similar to those reported in the handbook (for more details see Cornoldi et al., 2005). For the purposes of this study, a distinction between good and poor strategies was validated by administering the SQ1 questionnaire to 38 teachers. All of them taught 11- to 15-year-old students, worked in north-east Italy and taught Mathematics, Italian, or a foreign language. The teachers rated the utility of each strategy on a scale from 1 to 4 (irrespective of how often they themselves used that strategy). They rated the good strategies higher than the less effective ones (good strategies: $M = 3.33$ $SE = 0.036$, less effective strategies $M = 1.73$ $SE = 0.051$), with

$t(37) = 24.66 \, p \leq .001$. Examples of good strategies would be "Before reading observe the titles, subtitles, the words highlighted, and the figures" and "While reading check comprehension", whereas examples of less good strategies would be "Read the text once aloud" and "While studying keep the TV on as background".

Study task. The study task (ST) is also taken from the Italian standardised battery AMOS 8–15 (Cornoldi et al., 2005). Unlike a reading comprehension task, students must not only understand the text but also memorise its content. In fact, the task involves learning a text (982 words) entitled "Limpopo Park", which describes the geographical characteristics of the fauna and flora of an African natural park. The procedure involved a study phase of 30 min, a delay of 15 min, followed by three recall tasks. The first task required choosing, from a list of eight potential titles, the three that best summarised the text content (titles); the second required answering six cued recall questions (open questions); and the third required answering fifteen true/false questions (true/false questions). Cronbach's alpha measured in the current sample of students for the total score is .73. This is substantially better than that for those of the normative sample .47 (for more details see Cornoldi et al., 2005). Performance in this task is positively correlated with school performance using teacher evaluations on study ability ($r = .45$, $p \leq .001$; Cornoldi et al., 2005).

Procedure

An expert administered the SQ1, SQ2, and ST to whole classes including the experimental subjects. The students were instructed to rate the knowledge of each study strategy, irrespective of their actual use in the SQ1 and to rate how much they reported using each study strategy independently of their knowledge in the SQ2. The exact instructions for SQ1 (Utility) were as follows: "This is a list of activities that could be used effectively for studying a text (a chapter or a paragraph). Read each one carefully and rate how effective it is for you irrespective of what you usually do." The exact instructions for SQ2 (Use) were as follows: "Think about your habitual activities during the study phase. Read this list of study activities carefully and rate to what extent you applied the behaviour described, irrespective of its usefulness." Instructions included an example of the strategy rating in both questionnaires. There was a delay of 30 min between administration of the SQ1 and the SQ2. When giving ST instructions the examiner specified the amount of time for study and stated that, during this period, it was possible to underline and make notes in the text. One example for each measure (titles, open questions, true/false questions) was provided. The order of administration of the tasks was: (1) SQ1; (2) studying the ST text; (3) SQ2; (4) the recall tasks of ST. This schedule was decided on the basis of the following considerations: In order to have a study/memory test rather than a reading comprehension test we needed a substantial time interval between study and recall tasks in which students were involved in other activities. Furthermore, administration of strategy questionnaires before and after the study phase made the students aware of their strategy knowledge before actually studying and after an actual study experience. Teachers responsible for coordinating the teacher team for that class rated the study skills of their students using a Likert scale from 1 (low study ability) to 4 (good study ability).

Student selection

The scores for the ST were assigned in accordance with the instructions in the handbook: 1 point for each title correctly selected (maximum score 3), from 0 to 2 points for each open question (maximum score 12), 1 or -1 points for each true/false sentence correctly or erroneously verified respectively (maximum score 15). The total ST score was obtained adding up the score for each measure. The first and third quartiles (25° and 75°) derived from the normative data were used to select the low and high study skills students. Accordingly we selected 82 students with total scores on the ST $\leq 25°$ percentile, who composed the low study skills group; 128 students with total scores on the ST $\geq 75°$ percentile, who composed the high study skills group. Both study skills groups are representative of all grades (see Table 1). The high study skill group scored better in the study task as well as in the teacher's evaluation of study ability than low study skill group: study task, $F(1, 208) = 758.18$, $p \leq .001$; teacher evaluation of study skills, $F(1, 208) = 92.34$, $p \leq .001$ (means and standard errors are presented in Table 2). Students with good study skills formed a larger group, suggesting that the overall abilities level of our initial sample was higher than that of the normative sample.

TABLE 1
Number and gender composition of the low and high study skill groups

Grade	Number	Low study skills	High study skills
12-year-old students	38	19 (13 M, 6F)	19 (12M, 7F)
13-year-old students	57	16 (13 M, 3F)	41 (37 M, 4F)
14-year-old students	23	6 (3 M, 3F)	17 (14 M, 3F)
15-year-old students	92	41 (31 M, 10F)	51 (21 M, 30F)
Total	210	82 (60 M, 22 F)	128 (84 M, 44 F)

TABLE 2
Means and standard errors of the objective study task scores and
teacher ratings of study ability in low and high study skill groups
(teacher ratings from 1 = "low study ability" to 4 = "high study ability")

	Low study skills	High study skills
Study task		
M	6.71	19.99
SE	0.39	0.29
Teacher evaluation on study ability		
M	2.00	3.18
SE	0.09	0.07

RESULTS

Strategy questionnaires

We calculated six strategic indexes: (1) Knowledge of good strategies, (2) Knowledge of less effective strategies, (3) Use of good strategies, (4) Use of less effective strategies, (5) Strategic consistency of good strategies, (6) Strategic consistency of less effective strategies. Knowledge (SQ1) and use (SQ2) scores were calculated separately but in the same way; both questionnaire rating scores were based on the sum of ratings for (1) good and (2) less effective strategies and the two sums were divided by the numbers of strategies involved, i.e., by 22 and 10, respectively. The strategic consistency scores were calculated by summing the absolute values of the differences between knowledge and use ratings for each strategy; here too, the sum was divided by 22 and 10, respectively. Absolute values were used because a lack of consistency might be due to a use rating lower than a knowledge rating, or vice versa. Consequently, a low score reflected a smaller difference between knowledge and use ratings, i.e., high strategic consistency, while high scores reflected a larger difference between knowledge and use ratings, i.e., low strategic consistency. Use of a difference score was validated and recommended in previous studies (see De Beni et al., 2003) and in this context appeared more appropriate than other indexes (see Moè et al., 2001).

The means and standard errors of knowledge ratings, use ratings and strategic consistency scores of good and less effective strategies in students with low and high study skills are presented in Table 3. For each index calculated from the strategy questionnaires (knowledge rating, use rating, and strategic consistency score) a 2 × 2 analysis of variance was performed comparing the mean ratings given to the good and less effective strategies by high and low study skills students. These analyses of variance included the

TABLE 3
Means and standard errors of the knowledge ratings, use ratings,
and strategic consistency scores of the good and less effective
strategies in low and high study skill groups

	Low study skills	High study skills
Knowledge ratings		
Less effective strategies		
M	2.36	2.19
SE	0.04	0.03
Good strategies		
M	2.81	2.93
SE	0.04	0.03
Use ratings		
Less effective strategies		
M	2.17	2.01
SE	0.04	0.03
Good strategies		
M	2.32	2.56
SE	0.05	0.03
Strategic consistency scores		
Less effective strategies		
M	0.61	0.46
SE	0.03	0.02
Good strategies		
M	0.79	0.65
SE	0.03	0.03

strategy goodness (good vs. less effective) as a within-participant factor and study skills (high vs. low) as a between-participants factor.

Judgement of knowledge

Results of the 2×2 ANOVA for mixed design (Groups $\times$ Strategy goodness) on knowledge ratings indicated a significant effect of strategy, $F(1, 208) = 349.47$, $MSE = 0.091$, $\eta^2 = .64$, $p \leq .001$, due to the fact that in general all students rated the good strategies ($M = 2.87$, $SE = 0.02$) higher than less effective strategies ($M = 2.27$, $SE = 0.02$).

The ANOVA also showed interaction between strategy and study skills, $F(1, 208) = 21.33$, $MSE = 0.091$, $\eta^2 = .099$, $p \leq .001$. Means and standard errors of the knowledge rating differentiated into good and less effective strategies on the basis of high and low study skills are presented in rows 1 and 2 of Table 3. We carried out planned comparisons analysing the differences in strategy ratings between high and low study skills groups. We found a significant difference between high and low study skills students both in the less effective strategies, $F(1, 208) = 12.02$, $p \leq .001$, and in the good strategies, $F(1, 208) = 5.66$, $p = .018$, due to the fact that low study skills students rated the less effective strategies ($M = 2.36$, $SE = 0.04$) more highly than high study skills students ($M = 2.19$, $SE = 0.03$), whereas they rated the good strategies ($M = 2.81$, $SE = 0.04$) lower than did the high study skills students ($M = 2.93$, $SE = 0.03$). To summarise, students with low study skills were able to discriminate between good strategies but they did it in a less clear cut way than students with high study skills giving lower ratings to good strategies and higher ones to inadequate strategies than did high study skills students.

Judgement of use

The results of the 2×2 ANOVA concerning use ratings showed a significant effect of strategy, $F(1, 208) = 108.46$, $MSE = 0.106$, $\eta^2 = .35$, $p \leq .001$. As previously found with knowledge, but with a general decrease in the emphasis in the use rating, all students rated good strategies ($M = 2.44$, $SE = 0.03$) higher than less effective strategies ($M = 2.09$, $SE = 0.02$).

The results showed an interaction between strategy and study skills, $F(1, 208) = 34.01$, $MSE = 0.106$, $\eta^2 = .15$, $p \leq .001$. Means and standard errors of the use rating differentiated into good and less effective strategies on the basis of high and low study skills are presented in rows 3 and 4 of Table 3. As for the knowledge ratings, planned comparisons confirmed significant differences between study skills groups both in less effective strategies, $F(1, 208) = 9.43$, $p = .002$, and in good strategies, $F(1, 208) = 18.36$, $p \leq .001$; in fact low study skills students rated less effective strategies ($M = 2.17$, $SE = 0.04$) higher than the high study skills students ($M = 2.01$, $SE = 0.03$) and they rated the good strategies ($M = 2.32$, $SE = 0.05$) lower than did the high study skills students ($M = 2.56$, $SE = 0.03$). These results confirmed our hypothesis where students with low study skills give a lower self-rating in the reported use of good strategies and a higher self-rating in the reported use of less effective strategies in comparison with students with high study skills.

Strategic consistency

Results of the 2×2 ANOVA concerning the strategic consistency score indicated a significant effect of strategy, $F(1, 208) = 57.80$, $MSE = 0.049$, $\eta^2 = .23$, $p \leq .001$. The mean score of strategic consistency for good strategies ($M = 0.72$, $SE = 0.02$) was higher than for less effective strategies ($M = 0.53$, $SE = 0.02$). In other words, there was greater discrepancy between knowledge and use ratings for good than for less effective strategies.

Furthermore, a significant effect of study skills was found, $F(1, 208) = 16.37$, $MSE = 0.12$, $\eta^2 = .08$, $p \leq .001$, due to the fact that the strategic consistency score was lower (high discrepancy between knowledge and use ratings) in the low study skills students ($M = 0.70$, $SE = 0.03$) than in high study skills students ($M = 0.55$, $SE = 0.02$).

In this case the interaction between strategy and study skills was not found ($F < 1$) confirming that students with poor study skills were less consistent for both types of strategies. In fact, detailed inspection of the results showed that the low study skills group showed lower strategic consistency than the high study skills group, both for the less effective, $F(1, 208) = 15.68$, $p \leq .001$, and for the good strategies, $F(1, 208) = 9.04$, $p = .003$: Low study skills students showed greater discrepancy between knowledge and use ratings both in less effective strategies ($M = 0.61$, $SE = 0.03$) and in good strategies ($M = 0.79$, $SE = 0.03$) than the high study skills students (less effective strategy: $M = 0.46$, $SE = 0.02$; good strategies: $M = 0.65$, $SE = 0.03$). Furthermore comparison between good and less effective strategy scores showed a significant difference both in low study skill students, $F(1, 81) = 25.29$, $p \leq .001$, and in high study skill students, $F(1, 127) = 38.60$, $p \leq .001$, who reveal a higher degree of consistency in less effective strategies than good strategies. Means and standard errors for the strategic consistency score differentiated into good and less effective strategies on the basis of high and low study skills are presented rows 5 and 6 of Table 3.

Analysis of single strategies

The means and standard errors of knowledge and use ratings and strategic consistency score for each strategy (22 good and 10 less effective) differentiated for high and low study skills groups are presented in Table 4. We reported probability for all comparisons, but, due to the high number of comparisons, we considered significant only differences with $p < .001$. Significant differences between high and low study skill groups in the indexes (knowledge, use, and strategic consistency) are highlighted in grey in the corresponding row (see Table 4).

Strategies with most evident differences concerned depth of processing. Results showed that for two strategies (one good and one less effective) there was a significant difference between two study skill groups in knowledge, use, and strategic consistency. For Strategy 18 ("After having studied the text repeat its content in your own words"), high study skills students presented higher scores in knowledge, use, and they were more consistent than low study skill students. For Strategy 10 ("Skip what you didn't understand") high study skills students gained lower scores in knowledge, use, and they were more consistent than low study skill students.

TABLE 4
Means and standard errors for each strategy (good and less effective) in knowledge ratings, use ratings, and strategic consistency score in high and low study skill groups

Strategies*	List of strategies	Knowledge			Use			Strategic consistency		
		Low study skills	High study skills	F	Low study skills	High study skills	F	Low study skills	High study skills	F
Good strategies										
1	Think about the concepts already known about the topic	3.02 (0.08)	2.87 (0.06)	F=2.32 p=.13	2.41 (0.09)	2.50 (0.08)	F=.41 p=.52	0.80 (0.09)	0.73 (0.06)	F=0.45 p=.50
2	Before reading observe the title, subtitles, the words highlighted, and the figures	2.69 (0.09)	2.66 (0.06)	F=0.91 p=.76	2.49 (0.11)	2.46 (0.07)	F=.06 p=.81	0.79 (0.07)	0.56 (0.05)	F=7.43 p≤.01
4	Decide how to study the text and/or organise the study activity (subdivision in parts, the time of the study, etc.)	2.60 (0.10)	2.88 (0.08)	F=5.06 p=.03	1.99 (0.10)	2.17 (0.08)	F=2.06 p=.15	0.97 (0.09)	0.80 (0.06)	F=2.37 p=.13
5	Scan the text before reading	2.48 (0.09)	2.54 (0.08)	F=0.23 p=.63	2.11 (0.09)	2.48 (0.18)	F=2.43 p=.12	0.67 (0.08)	0.85 (0.16)	F=0.73 p=.39
6	While reading try to foresee the subsequent contents	1.72 (0.08)	1.48 (0.06)	F=5.59 p=.02	1.60 (0.09)	1.34 (0.05)	F=7.24 p≤.01	0.65 (0.08)	0.49 (0.05)	F=4.79 p=.03
9	While reading check comprehension	3.01 (0.09)	3.41 (0.06)	F=16.14 p≤.001	2.62 (0.10)	3.32 (0.07)	F=38.74 p≤.001	0.75 (0.08)	0.44 (0.05)	F=12.09 p≤.01
12	Reread parts of the text if misunderstood	3.32 (0.16)	3.72 (0.04)	F=7.99 p≤.01	2.58 (0.10)	3.53 (0.04)	F=82.21 p≤.001	0.99 (0.15)	0.38 (0.05)	F=21.69 p≤.001
14	Underline and highlight the relevant information, after reading it at least once	3.08 (0.08)	3.35 (0.06)	F=7.37 p≤.01	2.78 (0.11)	2.86 (0.08)	F=0.33 p=.56	0.74 (0.08)	0.83 (0.07)	F=6.1 p=.43
16	Observe the figures and read the related captions	2.58 (0.11)	2.43 (0.06)	F=1.48 p=.23	2.28 (0.11)	2.28 (0.08)	F=0.01 p=.95	0.71 (0.09)	0.47 (0.05)	F=6.98 p≤.010
18	After having studied the text repeat its content in your own words	3.37 (0.08)	3.79 (0.05)	F=24.03 p≤.001	2.99 (0.10)	3.60 (0.06)	F=30.20 p≤.001	0.67 (0.08)	0.33 (0.05)	F=14.50 p≤.001

Table 4 (*Continued*)

Strategies*	List of strategies	Knowledge Low study skills	Knowledge High study skills	F	Use Low study skills	Use High study skills	F	Strategic consistency Low study skills	Strategic consistency High study skills	F
19	Try to memorise the main point using a trick (rhymes, associations, stories, etc.)	2.56 (0.11)	3.12 (0.25)	$F=2.81$ $p=.092$	2.13 (0.11)	2.42 (0.09)	$F=3.89$ $p=.06$	0.74 (0.08)	0.88 (0.25)	$F=0.18$ $p=-.67$
20	While studying write guiding concept next to the text on a sheet	2.93 (0.09)	3.23 (0.07)	$F=6.68$ $p\leq.01$	2.28 (0.11)	2.71 (0.09)	$F=9.40$ $p\leq.01$	0.93 (0.08)	0.68 (0.07)	$F=5.14$ $p=.024$
23	After having studying the text write a summary	2.47 (0.11)	2.30 (0.08)	$F=1.73$ $p=.18$	1.78 (0.10)	1.64 (0.08)	$F=1.27$ $p=.26$	0.84 (0.08)	0.77 (0.06)	$F=0.45$ $p=.51$
24	Read silently trying to understand	2.67 (0.10)	2.71 (0.08)	$F=0.15$ $p=.70$	2.30 (0.11)	2.80 (0.08)	$F=11.33$ $p\leq.01$	0.60 (0.08)	0.45 (0.05)	$F=3.19$ $p=.07$
25	Think of questions the teacher might ask	3.10 (0.08)	2.94 (0.07)	$F=2.41$ $p=.12$	2.56 (0.10)	2.63 (0.08)	$F=0.23$ $p=.63$	0.80 (0.09)	0.58 (0.05)	$F=5.51$ $p=.02$
26	After the study try to do schemas, diagrams, or tables	2.67 (0.11)	2.70 (0.08)	$F=0.80$ $p=.78$	1.93 (0.10)	2.17 (0.08)	$F=3.47$ $p=.064$	0.85 (0.09)	0.72 (0.06)	$F=1.69$ $p=.19$
27	Repeat after having finished study	3.07 (0.08)	3.19 (0.06)	$F=1.41$ $p=.24$	2.77 (0.10)	3.15 (0.08)	$F=8.59$ $p\leq.01$	0.67 (0.09)	0.50 (0.05)	$F=3.22$ $p=.07$
28	Repeat the material after a certain period of time	2.79 (0.08)	2.96 (0.08)	$F=1.79$ $p=.18$	2.57 (0.11)	2.80 (0.09)	$F=2.64$ $p=.11$	0.63 (0.08)	0.59 (0.06)	$F=0.23$ $p=.63$
29	Repeat the material with a friend	2.64 (0.09)	2.63 (0.08)	$F=0.20$ $p=.89$	1.91 (0.10)	1.92 (0.07)	$F=0.01$ $p=.94$	0.95 (0.09)	0.80 (0.06)	$F=2.32$ $p=.13$
30	Explore the argument more thoroughly using other sources and looking for other information	2.56 (0.10)	2.73 (0.08)	$F=1.89$ $p=.17$	1.61 (0.08)	1.75 (0.07)	$F=1.51$ $p=.22$	1.02 (0.10)	1.05 (0.07)	$F=0.35$ $p=.85$
31	Take time to review parts of the text that were not so well learned	3.07 (0.09)	3.55 (0.04)	$F=27.03$ $p\leq.001$	2.75 (0.11)	3.41 (0.06)	$F=33.86$ $p\leq.001$	0.76 (0.09)	0.47 (0.05)	$F=8.90$ $p\leq.01$
32	Simulate an examination (oral or written) imagining to be in the situation	3.01 (0.09)	3.13 (0.08)	$F=0.94$ $p=.32$	2.39 (0.10)	2.41 (0.08)	$F=0.03$ $p=.86$	0.84 (0.09)	0.81 (0.07)	$F=0.74$ $p=.78$

Table 4 (*Continued*)

Strategies*	List of strategies	Knowledge			Use			Strategic consistency		
		Low study skills	*High study skills*	*F*	*Low study skills*	*High study skills*	*F*	*Low study skills*	*High study skills*	*F*
Less effective strategies										
3	Read the text once and aloud	2.89 (0.09)	3.02 (0.08)	$F=1.18$ $p=.28$	2.66 (0.10)	2.73 (0.10)	$F=0.26$ $p=.61$	0.56 (0.07)	0.51 (0.06)	$F=0.26$ $p=.61$
7	While studying keep music on as background	1.92 (0.12)	1.27 (0.05)	$F=29.74$ $p\leq.001$	1.74 (0.12)	1.37 (0.06)	$F=9.14$ $p\leq.01$	0.47 (0.08)	0.23 (0.05)	$F=6.35$ $p\leq.01$
8	Underline and highlight the relevant information before reading the complete passage	3.10 (0.10)	3.41 (0.06)	$F=7.86$ $p\leq.01$	2.93 (0.11)	3.08 (0.08)	$F=1.23$ $p=.27$	0.60 (0.09)	0.53 (0.06)	$F=0.51$ $p=.48$
10	Skip what you didn't understand	1.81 (0.10)	1.32 (0.05)	$F=23.32$ $p\leq.001$	1.89 (0.11)	1.21 (0.04)	$F=42.33$ $p\leq.001$	0.64 (0.08)	0.29 (0.04)	$F=16.55$ $p\leq.001$
11	Pay attention to the words and the information on the text without considering the figures	2.54 (0.10)	2.60 (0.07)	$F=0.31$ $p=.58$	2.33 (0.10)	2.51 (0.08)	$F=2.02$ $p=.16$	0.79 (0.08)	0.53 (0.06)	$F=6.74$ $p\leq.01$
13	While studying keep the TV on as background	1.55 (0.10)	1.11 (0.03)	$F=23.74$ $p\leq.001$	1.84 (0.11)	1.32 (0.06)	$F=19.39$ $p\leq.001$	0.43 (0.08)	0.27 (0.05)	$F=4.43$ $p\leq.01$
15	Copy out in a workbook the most difficult parts	2.09 (0.10)	1.95 (0.06)	$F=1.71$ $p=.19$	1.59 (0.09)	1.33 (0.06)	$F=6.64$ $p=.011$	0.68 (0.08)	0.69 (0.06)	$F=0.01$ $p=.92$
17	Rereading the text more than once, out loud and with expression	2.79 (0.11)	2.83 (0.08)	$F=0.08$ $p=.78$	2.43 (0.11)	2.45 (0.10)	$F=0.03$ $p=.86$	0.80 (0.08)	0.67 (0.06)	$F=1.77$ $p=.18$
21	After finishing your study, try to repeat the text literally	1.91 (0.09)	1.56 (0.06)	$F=10.93$ $p\leq.001$	1.66 (0.09)	1.40 (0.06)	$F=6.80$ $p\leq.01$	0.64 (0.08)	0.37 (0.05)	$F=9.89$ $p\leq.01$
22	Rereading aloud the text at least once	2.85 (0.09)	2.90 (0.08)	$F=0.13$ $p=.71$	2.62 (0.11)	2.73 (0.09)	$F=0.53$ $p=.47$	0.62 (0.08)	0.56 (0.06)	$F=0.41$ $p=.52$

The grey colour highlights the significant statistical differences of strategies scores between the two study skill groups. *Numbering corresponds to the presentation order of QS1.

Other critical strategies concerned comprehension monitoring and again deep text elaboration. For three strategies (two good and one less effective) there was a significant difference in knowledge and use and a tendency in consistency index ($p = .08$). For Strategies 9 ("While reading check comprehension") and 31 ("Take time to review parts of the text that were not so well learned") high study skills students presented greater score in knowledge, use, and they tended to be more consistent than low study skill students. For Strategy 13 ("While studying keep the TV on as background") high study skills students gained lower scores in knowledge, use, and they tended to be more consistent than low study skill students. Furthermore in Strategy 12 ("Reread parts of the text if misunderstood") high study skills students tended to gain higher scores in knowledge, scored significantly in use, and were more consistent than low study skill students. Finally, in dysfunctional Strategies 21 ("After finishing your study try to repeat the text literally") and 7 ("While studying keep music on as background") high study skills students gained lower scores in knowledge, tended to gain lower scores in use, but higher scores in strategic consistency than low study skills students.

DISCUSSION AND CONCLUSIONS

Metacognitive literature has shown that poor achievement is related to a poor knowledge about the utility and use of strategies (e.g., Pintrich & Schunk, 1996; Pressley et al., 1995; Schunk & Zimmerman, 1998), but the nature of this relationship has still not been considered in all its aspects. The present paper aimed to gain some understanding of this relationship by considering an important aspect associated with achievement, i.e., the study skill ability, which seems particularly dependent on good method of study.

Until now no research had investigated the differences between knowledge about the utility, use, and correspondence (strategic consistency) of study strategies, nor had the case of adolescent students with different study skills received particular attention. In the present research, study skills were analysed in relation to reported knowledge about the utility and use of strategies, distinguished for their efficacy, and the nature of this relationship was investigated using a metacognitive index, the strategic consistency, recently introduced in university study success researches (e.g., Moè et al., 2001) that measure the students' correspondence in knowledge and use ratings. These strategic aspects were measured using two strategy questionnaires (QS1 and QS2) and the study skill was measured with an objectives study task (ST) all included in a new standardised study battery (Cornoldi et al., 2005).

The results of the present study showed that all students, both with high and low study skills, rated good strategies as more effective than inadequate strategies (although the distance in ratings given to good and poor strategies was lower in students than in the group of teachers examined in the preliminary control). However, significant interactions between strategy goodness and study skills in knowledge and use ratings highlighted the fact that in low study skills students these discriminative abilities are less developed in comparison to high study skills students. These interactions showed that low study skill students attributed lower ratings to good strategies and higher ratings to the inadequate strategies in comparison with high study skills students both for knowledge and use ratings. The overall analysis confirmed that successful students are able to recognise the utility of good strategies and the usefulness of less effective strategies; they also reported using good strategies and not using the inadequate ones. Analysis of single strategies confirmed this pattern of results (Significant differences were present for some strategies, but for most of the others descriptive statistics showed tendencies in the same direction.)

These results must be considered together with the fact that, in general, knowledge ratings were higher than use ratings for both study skills groups (see means in Table 4). Even if this is a general tendency (motivated by the fact that it should difficult to use all strategies considered useful at the same time) it is found less often in successful students.

The pattern of the results is further emphasised by the strategic consistency index, less adequate in low study skill students than in high study skills students. It should be noted that high study skills students were particularly consistent in the case of poor strategies; in fact they obtained the highest strategic consistency score for the less effective strategies (0.46). In other words, successful students were able to recognise the inadequate strategies and consistently reported not using them. For high study skills students the strategic consistency index for less effective strategies was better than for good strategies (0.65). In fact, these students were able to recognise the utility of the latter but did not necessarily report using them. Low study skill students were in general less consistent than high study skill students even if this consistency was greater for less effective (0.61) than for good strategies (0.79). In fact, analysis of single strategies showed that low study skills students had a lower strategic consistency score, i.e., greater discrepancy between reported knowledge about the utility and their use, than high study skills students, in most strategies, even if differences were not significant in all.

Results of the present study provide new evidence, important for its educational implications, about metacognitive knowledge in adolescents with good study skills and the analysis of consistency between knowledge and use ratings. Comparisons between students with high and low study skills confirmed that failure in study tasks could reflect poor knowledge and use of good strategies and low consistency between knowledge and use of study strategies, as previously shown for university students (i.e., Moè et al., 2001). In general poor study skills students gave less importance to strategies that focus on cognitive and metacognitive aspects, which mainly affect learning, like a deep elaboration (see Metcalfe, Kornell, & Son, 2007 this issue) and self-testing (see Keleman, Winningham, & Weaver, 2007 this issue; McDaniel, Anderson, Derbish, & Morrisette, 2007 this issue).

Detection of single strategies where there was a discrepancy in strategic consistency scores between the study skill groups may offer an important insight for education. In particular poor study skills students appeared poorly consistent in reading comprehension monitoring strategies; students with low study skills were less able to recognise the utility and apply strategies that monitor the reading comprehension phase, such as: "While reading check comprehension" (Strategy 9), "Reread parts of the text if misunderstood" (Strategy 12), "After having studied the text repeat its content in your own words" (Strategy 18), "Take time to review parts of the text that were not so well learned" (Strategy 31); they were less able to recognise the uselessness and reported applying "Skip what you didn't understand"(Strategy 10) and "After finishing your study, try to repeat the text literally" (Strategy 21) more than high study skills students. Furthermore, low study skills students tended to be less consistent in dysfunctional strategies related to study context; in fact, they presented a discrepancy between knowledge about the utility and reported use, producing a low consistency, in "While studying keep music on as background" (Strategy 7) and "While studying keep the TV on as background" (Strategy 13).

There seem to be several reasons for low consistency in unsuccessful students. In particular they have more difficulty in implementing good strategies even though they know that some are effective or, by contrast, they use good strategies without recognising their efficacy. Furthermore, these students have more difficulty in inhibiting the use of less effective strategies even if they are recognised as useful, or by contrast they attribute higher utility to less effective strategies without a corresponding use. This confused representation of the utility and use of strategies produces inconsistent behaviour in study activity.

A further explanation of the dissociation between efficacy and use in the low study skills group may be related to motivational aspects: the effective strategies are the most demanding and to apply them students must employ cognitive abilities (i.e., attention, memory) modulated by motivational aspects such as effort, interest and self-regulation (e.g., Cornoldi et al.,

2003). This explanation may be valid for all students but particularly for low study skill students.

This lack of correspondence between knowledge of effective strategies and their use reflects an inadequate metacognitive strategic profile that produces poor performance in study activity. In these students, study behavioural habits are not related to what is considered important or known and this produces a distance between the real self and the ideal self (Higgins, 1987; Markus & Nurius, 1986). Obviously, it must be noticed that knowledge and use of a strategy or, more in general, of other metacognitive processes does not assure that the process is efficaciously used. For example, as Rawson and Dunlosky (2007 this issue) have shown, self-evaluation can be critical in learning but students tend to overestimate their knowledge, also under favourable conditions.

Taken together our results have implications for the theoretical under-standing of strategic abilities of low study skills students as well as for educational practice. As already stressed in the literature (e.g., Clearly & Zimmerman, 2004; Weinstein, Husman, & Dierking, 2000) educational practice should actively stimulate students to compare the utility of strategies and their corresponding use in order to enhance their consistency.

It seems essential to promote both strategic awareness and the differential use of more and less effective strategies; in fact, researches with university students have found that, when training is focused on strategic consistency, students with study difficulties improve their strategic abilities and their study performance (De Beni & Moè, 1997). In particular, it seems important to start practising reading comprehension monitoring strategies; an initial phase of training could be focused on improving use of consistent reading comprehension strategies that are recognised as useful, i.e., check compre-hension with reading the content, and, if the content is not understood, reread it and/or review that part. Harmonious development of strategic awareness and self-monitoring abilities may be a winning approach to improve strategic abilities, with a positive influence on study performance and better results in school achievement.

Finally, it should be noted that good strategic abilities can influence study success, but it is obvious that the opposite can be also true, as students who are good at text processing and learning will probably develop more adequate study strategies (see Schneider & Pressley, 1997, for more detail). Students with better learning abilities could be more aware and better able to monitor good strategies during the different learning phases. Thus, it is possible that knowledge and use of strategies affect study, but the opposite also occurs, i.e., good text processing skills affect development of high metacognition. Obviously, study success is also affected by other variables such as intelligence (De Ribaupierre & Lecerf, 2006), interference control

(Elliot, Barrilleaux, & Cowan, 2006) working memory (Wilhelm & Oberauer, 2006), but these variables interact with metacognitive competence in producing high school achievement.

REFERENCES

Artzt, A., & Armour-Thomas, E. (1998). Mathematics teaching as problem solving: A framework for studying teacher metacognition underlying instructional practice in mathematics. *Instructional Science, 26*(1–2), 5–25.

Bandura, A. (1986). *Social foundations of thought and action: A social cognitive theory.* Englewood Cliffs, NJ: Prentice Hall.

Beishiuzen, J. J., & Stoutjesdijk, E. T. (1999). Study strategies in a computer assisted study environment. *Learning and Instruction, 9,* 281–301.

Bembenutty, H., & Zimmerman, B.J. (2003, April). *The relation of motivational beliefs and self-regulatory processes to homework completion and academic achievement.* Paper presented at the annual meeting of the American Educational Research Association, Chicago (pp. 21–25). Report-Research; Speech/Meeting Papers (150).

Butler, D., & Winne, P. (1995). Feedback and SRL: A theoretical synthesis. *Review of Educational Research, 65,* 245–281.

Clearly, T. J., & Zimmerman, B. (2004). Self-regulation empowerment program: A school-based program to enhance self-regulated and self-motivated cycles of students learning. *Psychology in the Schools, 41*(5), 537–549.

Cornoldi, C. (1995). *Metacognizione e apprendimento [Metacognition and learning].* Bologna, Italy: Il Mulino.

Cornoldi, C., De Beni, R., & Fioritto, M. C. (2003). The assessment of self-regulation in college students with and without academic difficulties. In T. E. Scruggs & M. A. Mastropieri (Eds.), *Advances in learning and behavioral disabilities: Vol. 16. Identification and assessment* (pp. 231–242). Oxford, UK: Elsevier/JAI Press.

Cornoldi, C., De Beni, R., Zamperlin, C., & Meneghetti, C. (2005). *AMOS 8–15. Strumenti di valutazione di abilità e motivazione allo studio per studenti dagli 8 ai 15 anni. [AMOS 8–15: Assessment of the study skills and motivation for students from 8- to 15-year-olds].* Trento, Italy: Erickson.

De Beni, R., & Moè, A. (1997). Difficoltà di studio. Un intervento metacognitivo con studenti universitari [Study difficulties: A metacognitive training with university students]. *Psicologia Clinica dello Sviluppo, 1*(3), 433–440.

De Beni, R., Moè, A., & Cornoldi, C. (2003). *AMOS. Abilità and motivazione allo studio: Prove di valutazione e orientamento [AMOS: Study skills and motivation: Orientation and assessment tasks].* Trento, Italy: Erickson.

De Ribaupierre, A., & Lecerf, T. (2006). Relationship between working memory and intelligence from developmental perspective: Convergent evidence from a neo-Piagetian and Psychometric approach. *European Journal of Cognitive Psychology, 18*(1), 109–137.

Eccles, J., Roeser, R., Wigfield, A., & Freedman-Doan, C. (1999). Academic and motivational pathways though middle childhood. In L. Balter & Tamis-Lemond (Eds.), *Child psychology: A handbook of contemporary issues* (pp. 287–317). Philadelphia: Psychology Press.

Elliott, E. M., Barrilleaux, K. M., & Cowan, N. (2006). Individual differences in the ability to avoid distracting sounds. *European Journal of Cognitive Psychology, 18*(1), 90–108.

Gadzella, B. M. (1995). Differences in processing information among psychology course grade groups. *Psychological Reports, 77,* 1312–1314.

Garner, R. (1990). When children and adults do not use learning strategies. *Review of Educational Research, 60,* 517–530.

Higgins, E. T. (1987). Self-discrepancy: A theory relating self and affect. *Psychological Review, 94,* 319–340.

Hofer, B. K., & Pintrich, P. R. (2004). *Personal epistemology: The psychology of beliefs about knowledge and knowing.* Hillsdale, NJ: Lawrence Erlbaum Associates, Inc.

Kelemen, W. L., Winningham, R. G., & Weaver, C. A., III. (2007). Repeated testing sessions and scholastic aptitude in college students' metacognitive accuracy. *European Journal of Cognitive Psychology, 19,* 689–717.

Kleijn, W. C., van der Ploeg, H. M., & Topman, R. M. (1994). Cognition, study habits, test anxiety, and academic performance. *Psychological Reports, 75,* 1219–1226.

March, E. J., Roediger, H. L., III, Bjork, R. A., & Ligon Bjork, E. (2005). *Negative consequences of testing.* Paper presented at the Psychonomic Society, 46th annual meeting, Toronto, Canada.

Markus, H., & Nurius, P. (1986). Possible selves. *The American Psychologist, 41,* 954–969.

McDaniel, M. A., Anderson, J. L., Derbish, M. H., & Morrisette, N. (2007). Testing the testing effect in the classroom. *European Journal of Cognitive Psychology, 19.*

Metcalfe, J., Kornell, N., & Son, L. K. (2007). A cognitive-science based programme to enhance study efficacy in a high- and low-risk setting. *European Journal of Cognitive Psychology, 19,* 743–768.

Moè, A., Cornoldi, C., & De Beni, R. (2001). Strategic incoherence and academic achievement. In T. E. Scruggs & M. A. Mastropieri (Eds.), *Advances in learning and behavioral disabilities: Vol. 15. Technological applications* (pp. 237–258). Oxford, UK: Elsevier/JAI Press.

Moè, A., Cornoldi, C., De Beni, R., & Veronese, L. (2004). How can a student's depressive attitude interfere with the use of good self-regulation skills? In T. E. Scruggs & M. A. Mastropieri (Eds.), *Advances in learning and behavioral disabilities: Vol. 17. Research in secondary schools* (pp. 207–220). Oxford, UK: Elsevier/JAI Press.

Moreland, J. L., Dansereau, D. F., & Chmielewski, T. L. (1997). Recall of descriptive information: The roles of presentation format, annotation strategy, and individual differences. *Contemporary Educational Psychology, 22*(4), 521–533.

Nolen, S. B. (1988). Reasons for studying: Motivational orientations and study strategies. *Cognition and Instruction, 5,* 269–287.

Nolen, S. B., & Flaladyna, T. M. (1990). Personal and environmental influences on students' beliefs about effective study strategies. *Contemporary Educational Psychology, 15,* 116–130.

Pintrich, P., Wolters, C., & Baxter, G. (2000). Assessing metacognition and self-regulated learning. In G. Schraw & J. C. Impara (Eds.), *Issues in the measurement of metacognition* (pp. 43–97). Lincoln, NE: Buros Institute of Mental Measurements.

Pintrich, P. R., & Schunk, D. H. (1996). *Motivation in education: Theory, research, and applications.* Englewood Cliffs, NJ: Prentice Hall Merrill.

Pressley, M., Woloshyn, V., Burkell, J., Cariglia-Bull, T., Lysynchuk, L., McGoldrick, J., et al. (1995). *Cognitive strategy instruction that REALLY improves children's academic performance* (2nd ed.). Cambridge, MA: Brookline.

Rawson, K. A., & Dunlosky, J. (2007). Improving students' self-evaluation of learning for key concepts in textbook materials. *European Journal of Cognitive Psychology, 19,* 559–579.

Ruban, L. M., McCoach, D. B., McGuire, J. M., & Reis, S. M. (2003). The differential impact of academic self-regulatory methods on academic achievement among university studentes with and without learning disabilities. *Journal of Learning Disabilities, 36*(3), 270–286.

Schneider, W., & Pressley, M. (1997). *Memory development between 2 and 20.* Hillsdale, NJ: Lawrence Erlbaum Associates, Inc.

Schraw, G. (1998). Promoting general metacognitive awareness. *Instructional Science, 26,* 113–125.

Schunk, D. H., & Zimmerman, B. J. (1998). *Developing self-regulated learners: From teaching to self-reflective practice.* New York: Guilford Press.

Stoynoff, S. (1997). Factors associated with international students' academic achievement. *Journal of Instructional Psychology, 24*(1), 56–68.

Van Etten, S., Pressley, M., & Freebern, G. (1998). An interview study of college freshmens' beliefs about their academic motivation. *European Journal of Educational Psychology, 13*, 105–130.

Van Meter, P., Yokoi, L., & Pressley, M. (1994). College students' theory of note taking derived from their perceptions of note-taking. *Journal of Educational Psychology, 86*, 323–338.

Weinstein, C. E., Husman, J., & Dierking, D. R. (2000). Self-regulation interventions with a focus on learning strategies. In M. Boekaerts, P. Pintrich, & M. Seidner (Eds.), *Self-regulation: Theory, research, and applications* (pp. 727–747). Orlando, FL: Academic Press.

Weinstein, C. E., & Mayer, R. E. (1986). The teaching of learning strategies. In M. Wittrock (Ed.), *Handbook of research on teaching* (pp. 315–327). New York: Macmillan.

Wilding, J. E., & Valentine, E. (1992). Factors predicting success and failure in the first-year examinations of medical and dental courses. *Applied Cognitive Psychology, 6*, 247–261.

Wilhelm, O., & Oberauer, K. (2006). Why are reasoning ability and working memory capacity related to mental speed? An investigation of stimulus–response compatibility in choice reaction time tasks. *European Journal of Cognitive Psychology, 18*(1), 18–50.

Wolters, C. A. (1998). Self-regulated learning and college students' regulation of motivation. *Journal of Educational Psychology, 90*, 224–235.

Wood, E., Motz, M., & Willoughby, T. (1998). Examining students' retrospective memories of strategy development. *Journal of Educational Psychology, 90*, 698–704.

EUROPEAN JOURNAL OF COGNITIVE PSYCHOLOGY
2007, 19 (4/5), 650–670

Psychology Press
Taylor & Francis Group

The role of test anxiety in absolute and relative metacomprehension accuracy

Michael T. Miesner and Ruth H. Maki

Texas Tech University, Lubbock, TX, USA

Test anxiety reduces test performance and decreases students' confidence. We examined the relationship of prediction and posttest confidence judgements for self and others with multiple choice and essay test performance. High test-anxious participants (as measured by the Test Anxiety Scale) had poorer test performance on both multiple choice and essay tests than did low test-anxious participants, and high test-anxious participants gave lower judgements for themselves but not for others. Absolute metacomprehension accuracy as measured by the difference between confidence and performance showed a large degree of overconfidence on essay tests and much less overconfidence on multiple choice tests, but this effect did not interact with test anxiety. However, relative metacomprehension accuracy, as measured by intrasubject correlations between judgements and performance across texts, was higher for high text-anxious participants than for low test-anxious participants. Test anxiety may serve as a cue for how well material has been learned for high test-anxious participants.

Metacomprehension is the ability to judge one's learning from text. Typically, students read short texts, make predictions about future performance, and take tests on the texts. The correlations between predictions and performance for each individual are used as the measure of relative metacomprehension accuracy. Such correlations are often quite low, although they are usually greater than chance. In some studies, posttest confidence judgements are also made, and these are almost always more accurate than predictions (Maki, 1998). Relative metacomprehension accuracy is important for education in that a student who can accurately judge which portions of material are well-learned and which are not can select appropriate material for further study. Not studying information that is poorly learned will result in poor test performance and continuing to study

Correspondence should be addressed to Ruth H. Maki, Department of Psychology, Texas Tech University, Lubbock, TX 79409-2051, USA; E-mail: ruth.maki@ttu.edu

Michael Miesner is now at Middle Tennessee State University.

© 2007 Psychology Press, an imprint of the Taylor & Francis Group, an Informa business
http://www.psypress.com/ecp   DOI: 10.1080/09541440701326196

already-learned information wastes time that could be devoted to learning material that is not yet learned.

Absolute metacomprehension accuracy has also been studied with predictions and posttest confidence judgements (Maki, Shields, Wheeler, & Zacchilli, 2005). With this measure, the exact match between judgements and performance is of interest. Comparing the level of judgements with the level of performance allows researchers to determine whether participants are underconfident or overconfident. In education, accurate absolute accuracy allows students to have correct expectations about their overall performance on tests, both before and after taking the tests.

Several researchers have searched for relationships between relative metacomprehension accuracy (the relationship between predictions and performance for text) and students' verbal abilities, but such relationships are usually not found (Lin, Moore, & Zabrucky, 2001; Maki & McGuire, 2002). In contrast, studies using absolute metacomprehension accuracy often yield large individual differences with low ability individuals showing overconfidence and high ability individuals showing underconfidence (Kruger & Dunning, 1999; Maki, Shields, et al., 2005). Thus, relative metacomprehension accuracy has been impervious to the identification of variables that predict individual differences, but absolute metacomprehension accuracy shows large differences. However, metacomprehension studies have typically investigated only verbal ability or skill differences as measured by performance on the criterion test. Other types of individual differences have received much less attention. The present study investigated the role of test anxiety in absolute and relative metacomprehension. Test anxiety affects the level of confidence as well as performance (Lusk, 1981), so it is an individual difference variable that may be important in metacomprehension.

Test anxiety has been shown to decrease test performance (e.g., Lusk, 1981; Sarason, 1984; Tobias & Everson, 1997), although the reason for this is under debate. There is evidence both for an interference hypothesis that attributes poor performance to excessive attention to distracting thoughts during the test (Sarason, 1984; Wine, 1971) and a deficit hypothesis that attributes poor performance to poor study skills and lower levels of learning (Wittmaier, 1972). Wine's (1971) summary of research on test anxiety concluded that high test-anxious students have a diminished ability to concentrate on the test because they focus on how poorly they are performing and pay attention to self-depreciating thoughts. Birenbaum (1994) compared task irrelevant thoughts while students were completing a mathematics test. The students were divided into high and low achievement groups based on their mathematics course grade from the previous semester. Task irrelevant thoughts were more closely related to poor performance among the high achievement, high test-anxious group, supporting the interference view of test anxiety. However, Birenbaum concluded that the low achieving, high test-anxious group had lower levels of learning. Thus, performance deficits due to test anxiety may result both from interference during the test and from poorer learning, but in different groups of students. Tobias (1985) argued that both interference and defective skills may result from limited cognitive processing capacity in highly anxious individuals. Extraneous thoughts would leave less capacity available to process the material on a test. Additionally, students who have low levels of skill and high test anxiety would have further reductions in performance because tasks would be more difficult for them and require more of their cognitive capacity to perform. In addition, these students would need to cope with the loss of capacity due to extraneous thoughts.

High test anxiety may also influence performance expectations. In separate studies, Elliot and McGregor (1999) and Herman (1990) showed that fear of failure is a predictor of test performance. Elliot and McGregor found that fear of performing poorly increased worrying about performance during tests which, in turn, reduced test performance. Therefore, we hypothesised that students who are high in test anxiety would give lower prediction and posttest judgements in a metacomprehension task than students who are low in test anxiety.

A study by Stöber and Esser (2001) supports the hypothesis that students who have high test anxiety will give lower judgements. Participants read 60 sentences describing future tasks that they would need to remember. They were asked to decide whether they would use internal memory (try to memorise) or external memory (write it on the calendar). High anxiety participants selected the use of external memory more often than did low anxiety participants, suggesting that they knew they had poor memories. High anxiety participants also rated their success with internal memory strategies as lower than did low anxiety participants.

Several studies have examined the accuracy of test predictions as a function of test anxiety. Lusk (1981) asked students in a Psychology course to predict their future examination performance and to estimate their performance after the examination. She found that high test-anxious students gave lower prediction and posttest estimates than did students with low test anxiety. Lusk also measured absolute accuracy in that she calculated the difference between predictions and performance. Generally, she reported that test anxiety did not predict differences in prediction or posttest accuracy, but there was an effect of test anxiety for females on one examination. Lusk reported that the difference between predictions and performance was greater for female students who were lower in test anxiety than for female students who were higher in test anxiety. In this one condition, the high test-anxiety group was actually more accurate because they showed less overconfidence than the low test-anxiety group. Thus, Lusk produced solid evidence that students who are high in test anxiety give lower predictions and posttest estimates, and she provided some intriguing data that test anxiety may improve absolute prediction accuracy under some conditions.

Everson, Smodlaka, and Tobias (1994) produced some evidence that test anxiety may reduce the relative accuracy of predictions. They asked high and low test-anxious students to judge whether they knew the meanings of words or not. After students completed a vocabulary test with the words, each word was classified as matching the prediction or not. Students who were high in test anxiety had fewer matches than students who were lower in test anxiety. Tobias and Everson (1997) reported a similar study, except that mathematical equations were used rather than vocabulary words. They also found that students who had high test anxiety predicted less accurately than students who had lower test anxiety. However, Tobias and Everson also reported two other studies in which academically less able high-school students predicted vocabulary test performance. There was no relationship between the congruency of predictions and performance and test anxiety in either of those studies. They concluded that the high-school students in the last two studies may not have believed that poor performance would reflect negatively upon themselves and, thus, students with high trait test anxiety may not have been anxious in the experimental context. At any rate, the results with these tasks that involved prediction of performance over single items (vocabulary words or maths problems) produced mixed results about the relationship between test anxiety and prediction accuracy.

Tobias and Everson's (1997) tasks did not involve any learning in the experimental setting, so their results may not be directly relevant to our metacomprehension task in which participants learn, predict, take a test, and judge their performance. If test anxiety reduces cognitive capacity, then we might expect to see lower metacomprehension accuracy in students who have high test anxiety because monitoring learning involves additional cognitive capacity relative to reading alone (Rawson, Dunlosky, & Thiede, 2000). Extraneous thoughts would interfere with participants' abilities to read, comprehend, and monitor their levels of comprehension.

An alternative view is that test-anxious students have additional cues about how well they have learned. That is, they may be able to judge their levels of anxiety either during learning or on the test, and this may help them predict their future performance and judge their test performance. Dunlosky, Rawson, and Hacker (2002) proposed a levels-of-disruption hypothesis that suggests that disruptions in reading fluency can be used as cues to improve the accuracy of relative metacomprehension. Extraneous thoughts and other manifestations of anxiety may serve as cues to disruption. If such cues are diagnostic of future examination performance and if highly anxious students can use disruptions to make judgements about future performance, highly test-anxious students may be able to predict future performance better than less test-anxious students. Although highly test-anxious students perform more poorly on the criterion test, they may be able to judge their relative performance on various portions of their learning more accurately than students who are low in test anxiety. Furthermore, anxiety during the test itself may serve as a cue for posttest confidence judgements, increasing test-anxious students' posttest judgement accuracy relative to judgements made by less test-anxious students. In terms of absolute metacomprehension accuracy, it is expected that highly anxious students will reduce their predictions, which should reduce their level of overconfidence. This may benefit their absolute accuracy because the highly test-anxious students are the ones who do most poorly and who should give lower judgements. Lusk's (1981) finding of greater accuracy with increasing test anxiety in female participants supports this hypothesis.

The amount of decrement from test anxiety on performance and participants' metacognitive accuracy may depend upon the type of test that students are taking. Choi (1998) studied the relationship between test format and test anxiety. Two sections of a class were assigned to take essay tests and two sections were assigned to multiple choice tests. Multiple choice tests were perceived by students as being easier than essay tests, and, possibly as a result, students reported significantly less test anxiety when taking multiple choice than when taking essay tests. Choi also found greater test anxiety on essay tests among students who had negative attitudes towards essay tests than among students who had positive attitudes. However, differences in attitudes towards multiple choice tests did not predict test anxiety on multiple choice tests. Birenbaum and Feldman (1998) measured students' attitudes towards both open-ended (short answer essay) and multiple choice tests. They found that academically poorer students preferred multiple choice tests, and that students who were high in test anxiety were more likely to have negative attitudes towards open-ended formats than students who were low in test anxiety.

We used both multiple choice and essay tests, and we predicted differences in the accuracy of metacomprehension for the two test types. Several lines of evidence suggest that college students are poor at judging the accuracy of their answers to open-ended formats. Maki, Dempsey, Pietan, Miesner, and Arduengo (2005) reported that relative metacomprehension accuracy was greater with multiple choice tests than with essay tests, but this did not differ with verbal ability. Kang, McDermott, and Roediger (2007 this issue) reported other evidence suggesting that college students are poor at judging the accuracy of their essay answers. They found that difficult short answer tests that intervened between study and a final test boosted performance on the final test more than intervening multiple choice tests did, but only when corrective feedback was given. Feedback was particularly beneficial for questions that produced wrong answers on the intervening test. McDaniel, Anderson, Derbish, and Morrisette (2007 this issue) also found that short answer tests were more effective at boosting final test performance than multiple choice tests in their study, which incorporated feedback for open-ended answers. These results suggest that college students are not very good at judging their own answers to open-format tests, and they need external feedback to make accurate judgements. Rawson and Dunlosky (2007, this issue) reported a similar effect in a study in which students judged the accuracy of definitions that they had written. Students frequently judged definitions to be correct or partially correct even when the definitions were totally incorrect. Such overconfidence continued to occur when students were given the correct definitions to use in making their judgements. These studies suggest that judgements about essay test are particularly difficult. In addition, Birenbaum and Feldman (1998) and Choi (1998) reported that essay tests are more anxiety provoking and less preferred than multiple choice tests. These findings led us to the hypothesis that test anxiety would affect metacomprehension accuracy more with essay tests than with multiple choice tests.

In addition to asking students to predict and judge their own performance, we asked them to make judgements about the performance of others. Allwood (1994) asked students to answer general knowledge questions and then rate their confidence in their own answers and the answers of others. Allwood found that participants' judgements of others' answers were higher and more overconfident than judgements of their own performance. Relative metacognitive accuracy was not significantly different for self and other judgements. This suggests that self and other judgements were similar except that participants added a constant to each judgement when the target was another person rather than oneself. If highly anxious participants have a low perception of their own performance in comparison to the performance of others, then we may expect them to give lower confidence ratings than the low anxious participants when rating themselves, but not when rating others.

To summarise, we investigated absolute and relative metacomprehension accuracy as a function of level of test anxiety and test format. Participants predicted their own performance and the performance of other students, and they estimated their own performance and the performance of other students after completing the test. We expected that test anxiety would affect the absolute level of predictions and confidence judgements, but we were mainly interested in the effect of test anxiety on the accuracy of metacomprehension.

METHOD

Materials

A questionnaire was created for this experiment to obtain ratings of participants' anxiety before the experiment.[1] Initial anxiety was measured by the question "How would you rate your anxiety right now, about the upcoming test?" (with 1 = "not at all anxious" to 7 = "very anxious"). Trait anxiety was measured by the Test Anxiety Scale (TAS; Sarason, 1978). There are 37 items to which participants respond "true" or "false". Examples of questions are "Thoughts of doing poorly interfere with my performance on tests", and "I start feeling very uneasy just before getting a test paper back". The total number of responses that are consistent with test anxiety determine the TAS score. Test–retest reliabilities above .80 have been reported (Sarason, 1978).

Six expository texts were used for the reading section of the experiment. Texts were selected from a timed reading series by Spargo (1989) from booklets that vary in levels of difficulty. All texts were approximately 400 words in length. The multiple choice questions were supplied by the same booklets that included the texts (Spargo, 1989), except that we added an incorrect alternative to each question so that the multiple choice questions all had four alternatives instead of only three alternatives as printed in the booklets. There were six multiple choice questions for each text. The essay questions were similar to the multiple choice questions with the stems rephrased to elicit essay type answers from participants. There were four essay questions for each text.

Procedure

Participants were tested in a classroom in groups that ranged from one to thirteen in size. Each participant completed three packets. In the first packet, the participants were first asked to complete the mandatory consent form followed by the aforementioned questionnaire and the TAS (Sarason, 1978). After completion, the first packet was collected and students were given the second packet. The instructions informed students that they would be reading six texts and taking either an essay test or a multiple choice test over each text. Thus, they were informed about the type of test they would be taking before reading. The second packet included the six texts and the

---

[1] This questionnaire also contained questions asking about students' preference for multiple choice or essay tests, the strength of that preference, a prediction about how well they were likely to do on a multiple choice test, and a prediction about how well they were likely to do on an essay test. We did not use the data from these questions in the present analyses.

pretest prediction question for each text. Participants were given 2 min 30 s to read each text.[2] After participants read each text, they made predictions about performance. Instructions on each prediction page reminded participants that they would be taking an essay or a multiple choice test. Self predictions were made to the query "What percentage correct do you think you would get on a test over this text?" and predictions of others' performance were made to "What percentage correct do you think other students would get on a test over this text?" Participants wrote a percentage estimate in response to each question. They then rated their anxiety in response to "How anxious would you rate yourself right now?" with 1 = "not at all anxious" to 7 = "very anxious". Participants were given 30 s to complete the self and other predictions and the anxiety rating after reading the text. After the six texts were read and the predictions and anxiety ratings were made, the second packet was collected and participants received the third packet. This packet included the test (either multiple choice or essay) for each text, posttest confidence judgements for self and others, and a posttest anxiety rating scale. Posttest percentage estimates for self and others and state anxiety were measured using the same questions as in the postreading phase of the experiment except that future tense was changed to past tense for the confidence judgements. After participants completed the test and posttest ratings, they were debriefed.

Participants

A total of 82 college students enrolled in an Introductory Psychology course at Texas Tech University participated for credit in their class. Two participants were dropped because they did not complete the test and posttest ratings by the end of the experimental session. The final sample consisted of 47 females and 33 males. A total of 36 were randomly assigned to multiple choice test format and 44 were assigned to essay format. All participants were given the Test Anxiety Scale (TAS; Sarason, 1978) so that they could be divided using a median split based on their TAS score. The criterion for being placed into groups was a score of 13 or above for the high anxiety group and 12 or below for the low anxiety group. The low anxiety group consisted of 38 individuals and the high anxiety group consisted of 42 individuals.

---

[2] This time limit set the reading rate at an average of 375 words per min. This should have been adequate time for most readers to read the texts because Just and Carpenter (1987, p. 433) reported that the average reading rate is 240 words per min. We asked several undergraduate students to read the texts before conducting the study, and they were able to read them in fewer than 2.5 min.

RESULTS

Anxiety during the reading and judgement task

Several measures of anxiety during the reading and rating task can be used to determine whether trait anxiety as measured by the TAS was related to state anxiety during the experiment. In the beginning of the experimental session, participants rated their initial state anxiety on a 7-point scale. Mean ratings were analysed in a 2 (test anxiety group) × 2 (test format) analysis of variance (ANOVA). Although the low test anxiety participants gave a somewhat lower rating ($M = 2.54$) than the high test anxiety participants ($M = 2.93$), this difference was not significant, $F(1, 76) = 1.26$, $MSE = 2.38$, $p = .265$. Participants also rated their anxiety on the same scale immediately after reading each text and immediately after taking the test on each text. Although only a single question was used to measure state anxiety each time, the ratings were reliable. Cronbach's alpha for the six ratings given before reading each text was .874, and Cronbach's alpha for the six ratings given after taking each test was .978. Mean ratings were analysed in a 2 (test anxiety group) × 2 (test format) × 2 (time of rating) mixed-design ANOVA. Pretest and posttest ratings did not differ significantly ($M = 3.27$ vs. 2.42), $F(1, 76) = 1.86$, $MSE = 14.87$, $p = .176$, and time of rating did not interact with test format or anxiety group, $Fs < 1$. However, participants in the low anxiety group gave lower ratings ($M = 2.08$) than did participants in the high anxiety group ($M = 3.61$), $F(1, 76) = 4.81$, $MSE = 19.11$, $p = .031$, suggesting that the division based on trait anxiety predicted differences in state anxiety during the experiment. Anxiety group did not interact with test format, however, $F(1, 76) = 1.44$, $MSE = 19.11$, $p = .233$.

Percentage correct

Essay questions were scored in half point increments for each part of the correct essay response. Two points were possible for each of the four questions over each text, making each text worth a total of eight possible points. The essay test questions were scored by two separate readers. The Pearson r correlation for the two readers' mean scores across participants was .75 ($p < .001$). Because the scoring reliability was somewhat low, we averaged the two scores for each text and used these as our scores in the essay condition. The six multiple choice questions for each text were scored as correct or incorrect, giving a maximum score of six points for each text. Test scores were converted to percentages for analysis.

Percentages correct were analysed in a 2 (test anxiety group) × 2 (test format) between-subjects ANOVA. Percentages correct for the anxiety groups on multiple choice and essay tests are presented in the first data column of Table 1. Consistent with other literature (e.g., Lusk, 1981; Sarason, 1984), participants who scored high in test anxiety (as measured by TAS) scored significantly lower ($M = 49.78\%$) than participants who scored low in test anxiety ($M = 59.84\%$), $F(1, 76) = 13.05$, $MSE = 151.09$, $p = .001$. Test format was also an important factor. Test-takers scored significantly lower on essay tests ($M = 35.80\%$) than on multiple choice tests ($M = 73.81\%$), $F(1, 76) = 186.19$, $MSE = 151.09$, $p < .001$. Although the difference between high and low anxious participants was greater for the essay test (12.19%) than for the multiple choice test (7.94%), the interaction between these two variables was not significant, $F < 1$. Low anxious students did better than high anxious students on both types of tests.

Prediction and posttest confidence absolute accuracy

Students' absolute prediction and posttest confidence judgement accuracy were analysed by comparing the level of predictions and posttest confidence to actual percentages correct. Mean percentages for actual performance, predictions, and posttest confidence judgements are shown in the left half of Table 1. These three measures were included in a 2 (test anxiety group) × 2 (test format) × 3 (measure: prediction percentage, confidence percentage, actual percentage) mixed-design ANOVA. Overall, percentages differed across measures, $F(2, 152) = 107.04$, $MSE = 115.79$, $p < .001$, percentages were higher for multiple choice than for essay tests, $F(1, 76) = 34.24$, $MSE = 156.35$, $p < .001$, and percentages were higher for low anxiety than for high anxiety participants, $F(1, 76) = 9.57$, $MSE = 156.35$, $p < .003$. However, the interactions of the between-subjects variables with the three measures are critical for determining whether accuracy differed across conditions.

TABLE 1

Mean percentage correct, percentage predictions for self and others, and percentage confidence judgements for self and others for high and low anxiety participants on multiple choice and essay tests (with standard errors of the mean in parentheses)

		Self (%)		Others (%)	
	Correct	Prediction	Confidence	Prediction	Confidence
Multiple choice					
High anxiety	69.84 (2.68)	76.62 (2.83)	71.37 (4.26)	79.52 (1.99)	75.62 (3.04)
Low anxiety	77.78 (3.17)	84.23 (3.34)	83.09 (5.04)	80.57 (2.29)	81.86 (3.50)
Essay					
High anxiety	29.71 (2.68)	75.30 (2.83)	64.06 (4.26)	80.99 (2.27)	71.95 (3.47)
Low anxiety	41.90 (2.56)	80.62 (2.70)	71.86 (4.07)	79.23 (1.89)	73.52 (2.89)

Measure interacted significantly with test format, $F(2, 152) = 59.86$, $MSE = 115.79$, $p < .001$. Planned comparisons for multiple choice tests showed that predictions were higher than performance, $F(1, 34) = 9.01$, $MSE = 170.13$, $p = .005$, but posttest confidence judgements did not differ significantly from actual performance, $F(1, 34) = 2.72$, $MSE = 150.38$, $p = .108$. For essay tests, both predictions and posttest confidence judgements were significantly higher than actual performance, $F(1, 42) = 337.27$, $MSE = 231.39$, $p < .001$, and $F(1, 42) = 126.86$, $MSE = 357.78$, $p < .001$, respectively. Participants were overconfident about their multiple choice performance before but not after taking the test, but they were overconfident in their essay performance both before and after taking the test. Measure did not enter into any significant interactions with test anxiety, $Fs < 1$, showing that high and low test-anxious participants were equally overconfident and especially over-confident on their essay performance.

Predictions and posttest confidence for self and others

Each participant gave a prediction and posttest confidence judgement for themselves and for others for each text. We were particularly interested in how high and low anxiety participants perceived themselves versus others. Participants' ratings of themselves and others were analysed in a 2 (test anxiety group) $\times$ 2 (test format) $\times$ 2 (self vs. other) $\times$ 2 (predictions vs. posttest confidence) mixed-design ANOVA. Surprisingly, average percentage judgements did not differ significantly for multiple choice ($M = 79.28\%$) and essay ($M = 74.69\%$) tests, $F(1, 76) = 3.16$, $MSE = 518.53$, $p = .079$, although there were large differences in performance for the two test formats. Students were not sensitive to the fact that multiple choice tests would produce much higher performance than essay tests. Overall, judgements of high and low anxiety groups ($M = 74.66\%$ vs. $M = 79.32\%$) did not differ significantly, $F(1, 76) = 3.26$, $MSE = 518.53$, $p = .075$.

Predictions and posttest confidence judgements for self and others for the test anxiety and test format groups are shown in the rightmost four columns of Table 1. Overall, judgements did not differ significantly for self ($M = 75.89\%$) and others ($M = 78.08\%$), $F(1, 76) = 3.72$, $MSE = 100.13$, $p = .057$, but self/others entered into several interactions. The interaction between self/others and anxiety group was significant, $F(1, 76) = 9.28$, $MSE = 100.13$, $p = .003$. For judgements about self, high anxiety partici-pants made significantly lower judgements ($M = 71.84\%$) than did low anxiety participants ($M = 79.95\%$), $F(1, 76) = 5.87$, $MSE = 436.66$, $p = .018$. However, judgements about others did not differ for high anxiety and low anxiety participants ($M = 77.48\%$ vs. $M = 78.68\%$), $F < 1$.

Overall, predictions were higher than posttest confidence judgements (79.65% vs. 74.33%), $F(1, 76) = 13.16$, $MSE = 167.66$, $p = .001$. This effect did not interact with anxiety group, $F(1, 76) = 1.24$, $MSE = 167.66$, $p = .268$, but time did interact with test format, $F(1, 76) = 4.17$, $MSE = 167.66$, $p = .045$. Participants decreased their judgements significantly from predictions to posttest for essay tests (78.85–70.53%), $F(1, 42) = 11.56$, $MSE = 262.81$, $p = .001$, but this change was just short of significance for multiple choice tests (80.44–78.12%), $F(1, 34) = 3.77$, $MSE = 50.13$, $p = .060$. Self/ others also interacted with time of judgement, $F(1, 76) = 5.36$, $MSE = 23.89$, $p = .023$. The higher percentage given to others than to self was significant for posttest confidence judgements (76.06% for other vs. 72.59% for self), $F(1, 76) = 5.49$, $MSE = 85.46$, $p = .022$; but not for predictions (80.10% for others vs. 79.19% for self), $F < 1$. Each of these effects was independent of test anxiety, $Fs < 1$.

Relative metacomprehension accuracy

Most of the cognitive literature on metamemory uses a relative measure of accuracy. Participants' performance across units (texts in this case) is correlated with predictions or posttest confidence judgements across those units. The most common measure of these intraindividual correlations is the nonparametric correlation coefficient, gamma. Nelson (1984) argued that this is the best measure because it is not influenced by level of performance and because it assumes an ordinal and not an interval rating scale. Gamma varies from $+1.0$ for a perfect positive relationship to -1.0 for a perfect negative relationship. Most studies of metacomprehension have used gamma correlations for each individual as the dependent variable (e.g., Maki, 1998; Rawson et al., 2000). We calculated prediction and posttest gammas for self predictions and self confidence judgements and performance for each participant across the six texts. These are shown in Table 2.

We conducted single-sample t-tests on the gammas for each group to determine whether they were significantly different from zero. Both prediction and posttest confidence judgement gammas for the high anxious group taking multiple choice tests were significantly greater than zero. Gammas on the multiple choice test for the low anxious group were not significantly greater than zero. The posttest gammas for the essay test for the both anxiety groups were significantly greater than zero, but prediction gammas for the essay test were not significantly greater than zero.

These gammas were entered into a 2 (test anxiety) × 2 (test format) × 2 (predictions vs. posttest confidence judgements) mixed-design ANOVA.

TABLE 2

Mean intrasubject gammas between predictions and performance and between confidence judgements and performance for high and low anxiety groups taking multiple choice and essay tests (with standard errors of the mean in parentheses)

| Anxiety level | Multiple choice tests | | Essay tests | |
	Predictions	Posttest	Predictions	Posttest
High	.576* (.063)	.460* (.102)	.105 (.109)	.253* (.089)
Low	.199 (.182)	.193 (.141)	.053 (.075)	.245* (.106)

*Significantly greater than 0, $p < .05$, with a single-sample t-test.

Gammas were higher on the multiple choice than on the essay test, $F(1, 71)^3 = 3.99$, $MSE = 0.299$, $p = .050$, and gammas were higher overall for high anxiety than for low anxiety participants, $F(1, 71) = 4.51$, $MSE = 0.299$, $p = .037$. There was a trend towards an interaction between test format and anxiety, $F(1, 71) = 3.02$, $MSE = 0.299$, $p = .086$. As can be seen by the pattern of significance shown in Table 2, the anxiety effects tended to be larger in the multiple choice than in the essay test condition. Test format interacted with time of judgement, $F(1, 71) = 4.36$, $MSE = 0.131$, $p = .040$. Predictions and posttest confidence judgements did not differ in accuracy for multiple choice tests, $F < 1$, but posttest confidence judgements were more accurate than predictions for essay tests, $F(1, 39) = 4.17$, $MSE = 0.129$, $p = .048$.

Correlations of TAS with dependent variables.   We analysed our data by using a median split to divide participants into groups that were high or low in test anxiety. In order to show effects with test anxiety as a continuous variable, we also conducted Pearson r correlations between TAS scores and the other measures used in this study. These correlations are shown in Table 3. Although we found evidence for poorer performance on tests in the high anxiety group than in the low anxiety group, the overall correlation between anxiety and test performance was low and not significantly different from zero. However, when we calculated the correlations separately for the multiple choice and the essay groups, we found a fairly large and significant negative correlation between TAS scores and percentage correct on the essay

---

[3] Gamma cannot be computed if the percentage correct is identical across all texts or if the prediction or confidence judgement percentages are identical. We were not able to compute prediction gammas for three participants, and we were not able to compute posttest confidence judgement gammas for three participants. A total of five participants were lost in the combined analysis of prediction and posttest confidence judgement gammas because one participant was missing both gammas and four other participants were missing one of the gammas.

TABLE 3
Between-subject correlations of the test anxiety scale with performance,
predictions, posttest confidence judgements, prediction gammas, and posttest
confidence judgement gammas for the entire sample and for each test condition
separately (with Ns in parentheses)

	% correct	Predictions	Posttest judgements	Prediction gammas	Posttest gammas
Entire	−.095	−.470[1]	−.361[1]	.194[2]	.129
sample	(N =80)	(N =80)	(N =80)	(N =77)	(N =77)
Multiple choice	−.188	−.554[1]	−.472[1]	.211	.161
group	(N =36)	(N =36)	(N =36)	(N =35)	(N =35)
Essay group	−.477[1]	−.446[1]	−.387[1]	.097	.080
	(N =44)	(N =44)	(N =44)	(N =42)	(N =42)

[1]Significantly different from 0, $p < .05$; [2]significantly different from 0, $p < .10$.

tests. There was also a negative correlation between anxiety and performance in the multiple choice group, but it was not significant. Students who were higher in test anxiety did give lower predictions and lower posttest confidence judgements than students who were lower in test anxiety. This was true for the entire sample and for each test group. The positive correlation between prediction gammas and test anxiety was marginally significant for the entire sample, but not for the individual test groups. However, the sizes of the positive correlations were similar for the entire sample and for the multiple choice group, but the latter was based on a smaller number of individuals and not significant. Posttest confidence judgement accuracy did not relate significantly to test anxiety.

The correlations for TAS showed a similar pattern as the median-split analyses. In comparison to students who were low in test anxiety, students who were high in test anxiety performed more poorly, especially on the essay tests; gave lower predictions and posttest confidence judgements; and tended to show a stronger relationship between their predictions and performance.

Correlations between postreading anxiety, judgements, and test perform-ance. One hypothesis for why students who were higher in test anxiety predicted their performance more accurately is that they were able to use anxiety-related cues to make their judgements. If so, there should be a relationship between an individual's anxiety reported after reading a text and predictions. If anxiety is a mediating factor, then there should also be a relationship between postreading anxiety and performance. We computed two gamma correlations for each individual, one between postreading anxiety and predictions across the six tests and one between postreading anxiety and performance across the six texts.

Seven of the students taking multiple choice tests showed no variance in their postreading anxiety ratings and one student showed no variance in prediction judgements. For the remaining 28 students taking multiple choice tests, the mean gamma correlation between postreading anxiety and predictions was − .636 and the mean gamma between postreading anxiety and performance was − .302. Both of these values were significantly less than zero with single-sample t-tests. Thus, participants gave lower predictions to texts that were related to the most anxiety, and they performed more poorly on those texts. In the essay test condition, 13 students gave invariant postreading anxiety ratings, 1 student gave invariant prediction judgements, and 1 student showed no variance in either measure. For the remaining 29 participants, the mean gamma correlation between postreading anxiety and predictions was − .507, which was significantly less than zero. The correlation between postreading anxiety and performance in the essay condition, however, was only − .065, which did not differ from zero.

These correlations suggest that students used their postreading anxiety as a guide to predicting performance in both the essay and multiple choice conditions. This was an effective strategy in the multiple choice condition, but not in the essay condition. However, the alternative hypothesis, that students used their predictions as a guide to judging their anxiety, cannot be ruled out.

DISCUSSION

Test performance

Our results with the median split analysis replicated other studies (e.g., Lusk, 1981; Sarason, 1984) showing that students who have high test anxiety perform less well than students who have low test anxiety. We expected that this might interact with test format because multiple choice tests produce less test anxiety than open-ended essay tests (Choi, 1998), but we did not find the interaction in the median split analysis. However, when we calculated correlations between TAS scores and performance across participants, we found a significant negative correlation for the essay test, but not for the multiple choice test. Thus, there was some suggestion that overall anxiety was more of a negative factor on essay tests than on multiple choice tests.

Predictions and posttest confidence judgements

As expected from earlier studies (e.g., Lusk, 1981; Stöber & Esser, 2001), the high test-anxiety group gave lower predictions and posttest confidence judgements for themselves than did the low test-anxiety group. In addition, the Pearson *r* correlations using test anxiety as a continuous variable also showed that students who were higher in test anxiety gave lower prediction and posttest ratings. However, the highly test-anxious students did not give lower judgements about others, so the lower self judgements of high test-anxious participants were not simply due a bias to give lower judgements in general. Students who have high test anxiety apparently believe that they will perform worse than students with lower test anxiety, and, in fact, they do perform worse on average, so their judgements are accurate in that sense.

Metacomprehension accuracy

The earlier literature has been mixed on the accuracy of metacognitive judgements as a function of high and low test anxiety. Generally, Lusk (1981) found that absolute predictions and posttest confidence judgements about actual classroom tests were equally accurate for high and low test-anxious groups. However, the high anxiety female students did show more accurate absolute predictions on one test than did low anxious female students. We did not replicate this effect. The differences between predictions and performance and between posttest confidence judgements and performance were about the same for students who were low or high in test anxiety.

Participants in our study were much more accurate in judging the levels of their multiple choice performance than the levels of their essay performance. Although they were somewhat overconfident in their predictions about multiple choice test performance, their posttest confidence judgements did not differ significantly from their actual multiple choice performance. Our participants were highly overconfident in their judgements about essay tests both before and after the test. This replicates Rawson and Dunlosky's (2007 this issue) finding that students cannot accurately judge the accuracy of their open-ended answers. High overconfidence on an open-ended test format also helps to explain why Kang et al. (2007 this issue) found that intervening short answer tests were not helpful in increasing final test performance unless feedback was given. Students apparently thought that wrong answers on the intervening short answer tests were correct. Generally, students are more anxious about essay tests (Birenbaum & Feldman, 1998; Choi, 1998), so we expected that test anxiety would play a larger role in confidence judgements with essay tests than with multiple choice tests. However, anxiety levels were unrelated to the degree of overconfidence.

We also examined relative metacomprehension accuracy; that is, how well participants' judgements about individual texts correlated with their performance on those texts. We found evidence that students could predict and judge performance better on multiple choice tests than on essay tests, replicating a result found by Maki, Dempsey et al. (2005).

Overall, students who were higher in test anxiety predicted and judged their performance *better* than students who were lower in test anxiety. There was no interaction with type of test, but the higher accuracy was particularly apparent in the multiple choice test condition. In the correlation analysis, prediction accuracy tended to correlate with TAS scores although this was marginally significant only with the entire sample and not when test groups were analysed separately. As with the median split analysis, the trend was that anxiety predicted accuracy for multiple choice tests but not for essay tests. Predictions and posttest confidence judgements about essay tests were very inaccurate, suggesting the possibility of floor effects in the essay test condition. Students seem to be quite poor at judging their essay performance and higher levels of anxiety did not help them.

The only earlier studies in which relative prediction accuracy has been studied (Everson et al., 1994; Tobias & Everson, 1997) produced mixed results. Predictions about word knowledge and predictions about the ability to solve mathematical equations were *less* accurate among participants with high test anxiety than among participants with low test anxiety in two studies (Everson et al., 1994; Tobias & Everson, 1997), but not in two other studies (Tobias & Everson, 1997). Tobias and Everson (1997) speculated that the differences among their results may have occurred because students may not have been anxious in the contexts in which accuracy differences did not occur. In our study, we know that the high anxiety group was more anxious than the low anxiety group after reading each text and taking the tests. Higher levels of postreading anxiety for texts were related both to lower predictions and to poorer performance for those texts, especially with multiple choice tests. This suggests that anxiety may have mediated both levels of predictions and levels of accuracy on multiple choice tests. In our relative metacomprehension task, then, anxiety apparently facilitated accurate judgements about which texts would produce higher performance and which texts would produce lower performance.

This positive benefit of anxiety on metacomprehension accuracy is in contrast to the negative effect that anxiety had on test performance. Opposite patterns of effects for test performance and for metacomprehension accuracy have been reported in earlier studies. Weaver and Bryant (1995) showed that easy texts produced better test performance than texts of average difficulty, but the average difficulty texts produced higher levels of metacomprehension accuracy. Maki, Foley, Kajer, Thompson, and Willert (1990) showed that texts with deleted letters were remembered more poorly than texts with intact letters but the texts with deleted letters produced higher levels of metacomprehension accuracy. Rawson and Dunlosky (2002) showed that greater coherence among sentences improved memory performance but tended to result in poorer metacomprehension accuracy. Test anxiety may be added to this list of factors that decreases performance while increasing the accuracy of relative metacomprehension.

More accurate relative metacomprehension among high anxiety partici- pants is consistent with the levels-of-disruption hypothesis (Dunlosky et al., 2002). Participants who have high test anxiety should have more interfering thoughts and focus more on the quality of their performance rather than the on reading and test-taking tasks (Sarason, 1984; Wine, 1971), although this may be more likely for high achieving than for low achieving individuals (Birenbaum, 1994). The amount of attention given to disrupting thoughts and the amount of attention given to the experimental tasks may have served as a cue for predictions and posttest confidence judgements for the high anxiety participants. Whereas, anxiety may lower test performance overall, it may also serve as a cue for students to discriminate which texts will be associated with better performance and which texts will be associated with poorer performance. We found evidence that participants' levels of postreading anxiety were related to their predictions and that these anxiety levels were also related to test performance in the multiple choice test condition. When anxiety is a valid cue for performance as in the multiple choice condition, it should increase metacomprehension accuracy. A study that explicitly examines interfering thoughts might show that the accuracy of relative metacomprehension is related to the number of disruptions from self-focused thoughts.

Educational implications

The fact the test anxiety reduces test performance is well documented and has important implications for education. Metacomprehension accuracy also has important implications for education. Understanding how test anxiety affects metacomprehension accuracy may help to provide methods for improving the test performance of students who have high test anxiety. Our results suggest that students who have high test anxiety predict lower performance for themselves than students with low test anxiety, but that those predictions match the level of performance equally for both groups. Both high and low test-anxious groups were overconfident in their test predictions to about the same degree. The overconfidence was particularly evident on essay tests. Therefore, both groups would stop studying before material is adequately learned and test scores would be lower than those expected by students, especially on essay tests. The advice to high and low test anxiety students should be the same; that is, "study until you think you know the material and then study some more to ensure that you know it".

Students should also know that their expectations for essay tests are likely to be much higher than their likely performance levels.

Greater relative metacomprehension accuracy by high test-anxious students is interesting. It may be possible for these students to estimate their levels of distraction while reading and to use those estimates to gauge how much to study material. Material that was read when distracting thoughts were most numerous should lead to poorer performance than material that was read when distracting thoughts were fewer. Whether students can accurately judge the level of distraction that occurs while reading a text and whether these judgements relate to performance is an open question. Sarason (1978) developed the Cognitive Interference Questionnaire that asks participants to report task-irrelevant thoughts (such as "I thought about how poorly I was doing", "I thought about how much time I had left", and "I thought about my level of ability") on the task they had just completed. Thus, this questionnaire specifically taps into how much students are worrying about their future performance. It may be the amount of worry about performance rather than anxiety per se that gives cues to increase prediction accuracy. Asking participants to respond to the Cognitive Interference Questionnaire after reading each text and then recommending that they use the number of distracting thoughts to guide their predictions about future performance might increase relative metacomprehension accuracy. If so, then it might also be possible to improve test performance by asking students to use their level of distracting thoughts to guide them in how much time they should spend rereading texts. This could increase test performance and reduce the deleterious effects of test anxiety.

The concept of test anxiety brings a number of issues to the study of metacomprehension that have not been considered in the past. Test-anxiety researchers and metacomprehension researchers may both benefit by using concepts from the other field to better understand their own research areas. Better understanding of metacomprehension and test anxiety could be used to improve students' monitoring of their learning and to increase their levels of learning.

REFERENCES

Allwood, C. M. (1994). Confidence in own and others' knowledge. *Scandinavian Journal of Psychology, 35*, 198–211.

Birenbaum, M. (1994). On the relationship between test anxiety and test performance. *Measurement and Evaluation in Counseling and Development, 27*, 293–301.

Birenbaum, M., & Feldman, R. A. (1998). Relationships between learning patterns and attitudes towards two assessment formats. *Educational Research, 40*, 90–98.

Choi, N. (1998). The effects of test format and locus of control on test anxiety. *Journal of College Student Development, 39*, 616–619.

Dunlosky, J., Rawson, K. A., & Hacker, D. J. (2002). Metacomprehension of science text: Investigating the levels-of-disruption hypothesis. In J. Otero, J. A. León, & A. C. Graesser (Eds.), *The psychology of science text comprehension* (pp. 255–279). Mahwah, NJ: Lawrence Erlbaum Associates, Inc.

Elliot, A. J., & McGregor, H. A. (1999). Test anxiety and the hierarchical model of approach and avoidance achievement motivation. *Journal of Personality and Social Psychology, 76*, 628–644.

Everson, H. T., Smodlaka, I., & Tobias, S. (1994). Exploring the relationship of test anxiety and metacognition on reading test performance: A cognitive analysis. *Anxiety, Stress, and Coping: An International Journal, 7*, 85–96.

Herman, W. E. (1990). Fear of failure as a distinctive personality trait measure of test anxiety. *Journal of Research and Development in Education, 23*, 180–185.

Just, M. A., & Carpenter, P. A. (1987). *The psychology of reading and language comprehension.* Newton, MA: Allyn & Bacon.

Kang, S. H. K., McDermott, K. B., & Roediger, H. L., III. (2007). Test format and corrective feedback modify the effect of testing on long-term retention. *European Journal of Cognitive Psychology, 19*, 528–558.

Kruger, J., & Dunning, D. (1999). Unskilled and unaware of it: How difficulties in recognizing one's own incompetence lead to inflated self-assessments. *Journal of Personality and Social Psychology, 77*, 1121–1134.

Lin, L.-M., Moore, D., & Zabrucky, K. M. (2001). An assessment of students' calibration of comprehension and calibration of performance using multiple measures. *Reading Psychology, 22*, 111–128.

Lusk, S. L. (1981). Test anxiety, level and accuracy of predicted performance. *Psychological Reports, 49*, 527–532.

Maki, R. H. (1998). Test predictions over text material. In D. J. Hacker, J. Dunlosky, & A. C. Graesser (Eds.), *Metacognition in educational theory and practice* (pp. 117–144). Mahwah, NJ: Lawrence Erlbaum Associates, Inc.

Maki, R. H., Dempsey, C., Pietan, A., Miesner, M., & Arduengo, J. (2005, November). *Metacomprehension for multiple-choice, essay, and recall tests.* Paper presented at the annual meeting of the Psychonomic Society, Toronto, Canada.

Maki, R. H., Foley, J. M., Kajer, W. K., Thompson, R. C., & Willert, M. G. (1990). Increased processing enhances calibration of comprehension. *Journal of Experimental Psychology: Learning, Memory, and Cognition, 16*, 609–616.

Maki, R. H., & McGuire, M. J. (2002). Metacognition for text: Implications for education. In T. J. Perfect & B. L. Schwartz (Eds.), *Applied metacognition* (pp. 39–67). Cambridge, UK: Cambridge University Press.

Maki, R. H., Shields, M., Wheeler, A. E., & Zacchilli, T. L. (2005). Individual differences in absolute and relative metacomprehension accuracy. *Journal of Educational Psychology, 97*, 723–731.

McDaniel, M. A., Anderson, J. L., Derbish, M. H., & Morrisette, N. (2007). Testing the testing effect in the classroom. *European Journal of Cognitive Psychology, 19*, 494–513.

Nelson, T. O. (1984). A comparison of current measures of the accuracy of feeling-of-knowing predictions. *Psychological Bulletin, 95*, 109–133.

Rawson, K. A., & Dunlosky, J. (2002). Are performance predictions for texts based on ease of processing? *Journal of Experimental Psychology: Learning, Memory, and Cognition, 28*, 69–80.

Rawson, K. A., & Dunlosky, J. (2007). Improving students' self-evaluation of learning for key concepts in textbook materials. *European Journal of Cognitive Psychology, 19*, 559–579.

Rawson, K. A., Dunlosky, J., & Theide, K. W. (2000). The rereading effect: Metacomprehension accuracy improves across reading trials. *Memory and Cognition, 28*, 1004–1010.

Sarason, I. G. (1978). The Test Anxiety Scale: Concept and research. In C. D. Spielberger & I. G. Sarason (Eds.), *Stress and anxiety* (Vol. 5, pp. 193–216). Washington, DC: Hemisphere Publishing Corporation.

Sarason, I. G. (1984). Stress, anxiety, and cognitive interference: Reactions to tests. *Journal of Personality and Social Psychology, 46,* 929–938.

Spargo, E. (1989). *Timed readings* (3rd ed., Books 5, 7, & 10). Providence, RI: Jamestown Publishers.

Stöber, J., & Esser, K. B. (2001). Test anxiety and metamemory: General preference for external over internal information storage. *Personality and Individual Differences, 30,* 775–781.

Tobias, S. (1985). Test anxiety: Interference, defective skills, and cognitive capacity. *The Educational Psychologist, 20,* 135–142.

Tobias, S., & Everson, H. T. (1997). Studying the relationship between affective and metacognitive variables. *Anxiety, Stress, and Coping, 10,* 59–81.

Weaver, C. A., III, & Bryant, D. S. (1995). Monitoring of comprehension: The role of text difficulty in metamemory for narrative and expository text. *Memory and Cognition, 23,* 12–22.

Wine, J. (1971). Test anxiety and direction of attention. *Psychological Bulletin, 76,* 92–104.

Wittmaier, B. (1972). Test anxiety and study habits. *Journal of Educational Research, 65,* 852–854.

EUROPEAN JOURNAL OF COGNITIVE PSYCHOLOGY
2007, 19 (4/5), 671–688

Improving metacomprehension accuracy and self-regulation in cognitive skill acquisition: The effect of learner expertise

Anique B. H. de Bruin, Remy M. J. P. Rikers, and
Henk G. Schmidt

*Department of Psychology, Erasmus University Rotterdam, Rotterdam, The
Netherlands*

The positive learning effect of metacognitive strategy instructions based on the cue-utilisation framework (Koriat, 1997) has been shown in memorising word pairs and studying expository text. The present study explored the relevance of this framework in cognitive skill acquisition for learners with and without prior domain knowledge. That is, the quality of metacomprehension and self-regulation of novices and more experienced chess players were compared, when learning an endgame of chess. The experienced chess players not only became more skilled at the endgame, but also exemplified higher metacomprehension accuracy and self-regulation than the novices. The absolute level of metacomprehension accuracy and self-regulation of the novices was close to zero. This is the first study that shows an effect of expertise on metacomprehension accuracy in cognitive skill acquisition. Differences between cognitive skill acquisition and text comprehension that might explain differences in quality of metacognitive processes between these domains are discussed.

Over the past two decades, research on metacognition has been directed at uncovering the strategies self-regulated learners apply when approaching study tasks. For example, self-regulated learners set clear and attainable goals, monitor learning activities, and adjust these when deviations from attainment of these goals are identified (Winne, 1995, 2001; Zimmerman & Schunk, 2001). Theoretical models that attempt to explain self-regulated study can be described as discrepancy-reduction models (e.g., Butler & Winne, 1995; Dunlosky & Hertzog, 1997; Thiede & Dunlosky, 1999). That is,

Correspondence should be addressed to A. B. H. de Bruin, Department of Psychology, Erasmus University Rotterdam, T13–39, PO Box 1738, 3000 DR Rotterdam, The Netherlands. E-mail: debruin@fsw.eur.nl

The authors would like to thank Nikky van Dorp for her help in collecting the data. This research was supported by a grant from the Erasmus University Trust Fund.

the individual has a desired level of understanding of the to-be-studied material, and monitors to what extent this state has been reached. If a discrepancy exists between desired and current level of understanding, the learner will continue studying the material. During restudy, the learner will again monitor the level of understanding and determine whether further restudy is necessary. This process continues until the discrepancy between desired and current level of understanding has disappeared.

Evidence for this explanation is mainly found in research on memorising word pairs. Son and Metcalfe (2000) list a number of studies that demonstrate that if a learner is able to determine which material is understood, and which is not, more study time will be allocated to the least-understood material (for specific circumstances under which learners select the easiest items, see Dunlosky & Thiede, 2004; Metcalfe & Cornell, 2003). Moreover, Thiede (1999) showed that the level of monitoring accuracy determines the effectiveness of self-regulation of study. Learners who more accurately monitored learning, selected material that needed restudying more appropriately, and performed better on a subsequent criterion test. Note that Koriat, Ma'ayan, and Nussinson (2006) recently presented data demonstrating that monitoring not only affects control of study, but the reversed, control affecting monitoring, is also found.

In the word-pair paradigm, participants are typically asked to study a list of translation pairs (i.e., Swahili-English translations; Nelson, Dunlosky, Graf, & Narens, 1990). Afterwards, they are presented with the first word of each pair and asked to estimate the chance of retrieving the second word on a subsequent test. The higher these "judgements of learning" (JOLs) are, the more confident the learner is to retrieve the correct word on the test. Accurate monitoring is indicated by a highly positive correlation between JOLs and test performance. The learner was able to assign high JOLs to words that were indeed recalled, and low JOLs to words that were not recalled. After the JOL procedure, learners are asked to select word pairs they wish to restudy. Adequate self-regulation is characterised by a highly negative correlation between JOLs and selection of word pairs for restudy. That is, word pairs that were assigned high JOLs are not selected for restudy, whereas word pairs with low JOLs are selected. Note that adequate self-regulation of study is dependent on accurate monitoring: If learners assign incorrect JOLs to word pairs (e.g., high JOLs to word pairs that were not retrieved), they will not select the word pairs that in fact need restudying.

The positive relation between JOLs and subsequent performance, and the negative relation between JOLs and item selection, has been repeatedly established using the word-pair paradigm (e.g., Cull & Zechmeister, 1994; Dunlosky & Hertzog, 1997; Nelson et al., 1994; Thiede & Dunlosky, 1999). Koriat (1997) argues that, when providing JOLs, learners use certain cues that are indicative of ease or difficulty of learning. Within this cue utilisation view, learners are thought to apply specific rules or heuristics to arrive at a probability of recalling the item later. Although the cue utilisation approach has provided insight into the cognitive mechanisms that explain how JOLs predict memory performance, research in this field has generally led to few educational implications. Given that the relevance of memorising word pairs only applies to a small subset of the educational contexts, research is needed that addresses the validity of the cue utilisation framework in educationally more relevant situations. For example, research on studying expository text has shown that, when preceded by a delayed keyword or delayed summarisation instruction, learners are able to accurately assess metacomprehension[1] of the texts, and regulate further study behaviour (Thiede & Anderson, 2003; Thiede, Anderson, & Therriault, 2003; Thiede, Dunlosky, Griffin, & Wiley, 2005). It is crucial that the keyword or summarisation instruction is given at a delay after reading the text, to ensure that short-term memory has been cleared. When short-term memory for the text is cleared, generating keywords or providing a summary will access the situation model of the text, and will therefore provide a more accurate prediction of long-term recall.

A further step that until now has not been taken would be to test the applicability of the cue utilisation view in cognitive skill acquisition. Research on the effect of metacognitive strategies in cognitive skill acquisition could lead to educational implications that are transferable to other skill domains. Moreover, a second theme that merits attention in research on metacognition concerns the influence of learner expertise on metacognitive strategies as self-monitoring and self-regulation. More specifically, the question is to what extent learners of low versus high prior knowledge differentially benefit from training metacomprehension and self-regulation. There are, however, certain characteristics of novices that could impede self-regulated learning. For example, due to the absence of automated subskills, execution of the skill poses a high demand on working memory (Winne, 1995). Requiring novices to self-monitor and self-regulate their learning could therefore induce a working memory overload. Winne (1995) argues that self-regulated learning should be delayed until individuals have proceduralised the domain rules. On the other hand, evidence exists that metacognitive strategies can be taught effectively to inexperienced learners. Azevedo and Cromley (2004) provided participants with a 30-min training on how to use self-regulated learning before studying the circulatory

---

[1] The term used to denote the predictive accuracy of comprehension ratings in learning from text is "metacomprehension accuracy", whereas the term used in learning word pairs is "monitoring accuracy" (Maki, 1998). In chess, we also prefer the term "metacomprehension accuracy", since this more clearly emphasises that learners reflect on how well study material is understood.

system. Despite the low level of prior domain knowledge, those who received the self-regulated learning training experienced a larger shift in their mental models than those who did not. A study by Butler (1998) also revealed that self-regulation skills of poor readers could be improved through extensive training. Students were even able to transfer the learned skills to different contexts. Given the extensiveness of the metacognitive training these studies used, identifying facets of the training that caused the learning effect is difficult, and formulating educational implications is not possible at this point.

Although the abovementioned studies did not examine differences between novices and more experienced learners in metacognitive skills, there are a few studies that did (e.g., Eteläpelto, 1993; Prins, Veenman, & Elshout, 2006; Schneider, Schlagmüller, & Vise, 1998). For example, Eteläpelto (1993) showed that expert programmers possessed superior metacognitive knowledge compared to novices. The experts were more aware of the ideal working strategy to approach a problem and could better monitor their own working strategy. A study by Schneider et al. (1998) revealed that among 7- and 8-year-old soccer experts and novices general metacognitive knowledge contributed to recall of the soccer information for experts, but not for novices. Finally, Prins et al. (2006) showed that metacognitive skilfulness was correlated with quality of inductive problem solving for easy problems in novices and for intermediate problems in advanced learners. However, these studies all examined the spontaneous use of metacognitive skills by more and less experienced learners, but did not examine whether explicitly stimulating these groups to self-monitor and self-regulate would have a positive effect on learning. Therefore, we used the cue utilisation framework and existing research on metacomprehension accuracy to design a study that would ask learners to self-monitor and self-regulate and that would provide a quantitative measure of these processes allowing comparison between groups of varying expertise. The goal of this study was twofold. First, we assessed to what extent low prior knowledge learners are able to benefit from self-regulatory strategy training in chess skill acquisition. Second, we examined whether prior knowledge positively affects metacomprehension accuracy and self-regulation. Therefore, we compared novices and more experienced chess players on two specific forms of metacognitive activities, namely metacomprehension accuracy and self-regulation when studying the endgame of King and Queen against King (KQK endgame). Metacomprehension accuracy was calculated by correlating the quality of the prediction of a chess move with the JOL score that the learner assigned to this prediction. Self-regulation was computed by correlating the JOL score with the learner's decision ("yes" = 1, "no" = 0) to study another example of the same move. Given that the more experienced chess players had proceduralised the rules of the domain, we hypothesised that the more experienced chess players would not only learn to play the endgame better, but would also more accurately monitor understanding and regulate their study behaviour. We expected relatively low levels of metacomprehension and self-regulation in the inexperienced learners, due to a lack of proceduralised subskills.

METHOD

Participants

Participants were 38 undergraduate students from Erasmus University Rotterdam, The Netherlands (mean age = 20.5 years, $SD = 1.8$). Eighteen participants had never played chess before (i.e., "inexperienced chess players"[2]). The remaining 20 participants (i.e., "experienced chess players") had in previous years played chess, but infrequently. The selection of the experienced chess players consisted of three criteria. First, participants had a perfect score on a 9-item true–false test on the basic chess rules. More information about the basic chess rules is given below. Second, participants had to play three examples of the endgame of King and Queen against King (the KQK endgame) against a chess computer. Of these three endgames, participants had to be able to solve the easiest one, which was an example of a game of checkmate in one move. Of the remaining two (checkmate in three and nine moves) participants had to solve exactly one, but never in the minimum number of moves, to guarantee that they did not master the endgame. Finally, participants were asked about their chess activities in the past. Only participants who had never joined a chess club were allowed to take part in the experiment. Of the 107 participants tested, 20 met all three criteria and took part in the study.

Materials

At the start of the experiment, the inexperienced chess players studied a computer presentation about the basic chess rules (i.e., what are the legal moves for King and Queen? What is capturing, check, checkmate, and stalemate?). These rules can be distinguished from the principles that underlie the KQK endgame. Knowing what the legal moves of a Queen and King are does not provide enough information to play this endgame correctly. This chess game was selected because it could be studied in isolation without the need to explain all the chess pieces and rules. The chess

---

[2] This group is the same group as the "JOL forced selection condition" in de Bruin, Rikers, and Schmidt (2005).

computer program "Fritz 7" (World Chess Federation rating of around 2700) provided the chess exercises for the learning and test phase. All learning and test exercises were endgames of checkmate in six moves, representing the same underlying chess principles.

Procedure

The experiment consisted of three phases: the basic chess rules phase, the learning phase, and the test phase. In the basic chess rules phase, the inexperienced participants studied a presentation about the chess rules that were relevant for this endgame in a self-paced way. This took on average 25 min. Afterwards, participants completed the same true–false test that the experienced players had taken in the selection procedure, to assess whether they had understood the rules. Any errors were explained by the experimenter before proceeding to the learning phase. Given that the inexperienced learners were trained to use the rules in a declarative manner (i.e., by recognising situations in which the rules did or did not apply) and had never applied the rules in a problem situation, they were able to use them in a nonautomatic way, and it was unlikely that they had procedur-alised them (Anderson, 1983).

The structure of the learning phase was identical for both groups, and is presented in Table 1. The goal of this phase was to learn the KQK endgame, and at the same time monitor comprehension and self-regulate study behaviour. The learning phase consisted of five parts: the prediction phase (P in Table 1), the JOL phase (J), the reprediction phase (R), the move selection phase (S), and the restudy phase (RS). Because in this endgame Black only has a King, White (i.e., King and Queen) is, in theory, able to win. The best result possible for Black is a draw. White's task is to prevent a draw and win the endgame by correctly applying the endgame principles. These principles were nowhere stated, but had to be inferred by predicting and studying the computer moves. An example of a principle is that the King can only be checkmated at the edge of the board.

Participants started the learning phase by studying a KQK endgame and predicting the six moves for White. This prediction phase (P1, P2, ... P6 in Table 1) had the following structure:

1. Participants saw the start position and had 40 s to predict move one (P1) for White.
2. The computer played the optimal move for White. At this point, participants could compare their prediction to the computer move.
3. The computer played the optimal move for Black.

TABLE 1
Overview of the experimental procedure of the learning phase

Example	Phase	Task	Moves
	1. Prediction phase	Predict the next move of the computer for white.	P1, P2, ... P6
	2. JOL phase	How confident are you that you will correctly predict a similar move in the future? 0%–20% ... 80%–100%	J1, J2, ... J6
	3. Reprediction phase	Repredict the next move of the computer for white.	R1, R2, ... R6
	4. Selection phase	Would you like to study another example of this move (yes = 1, no = 0)	S1, S2, ... S6

Table 1 (*Continued*)

Example	Phase	Task	Moves
	5. Restudy phase	Predict the next move of the computer for white.	RS1, RS2, ... RS6

P1 = participants predicted move 1; J1 = Participants judged comprehension of move 1; R1 = Participants repredicted move 1; S1 = Participants selected move 1 for restudy; RS1 = Participants restudied move 1. Participants studied these five phases for each chess exercise. A total of four different chess exercises were studied.

This procedure was repeated for move 2 (P2) until move 6 (P6).

To promote thorough processing, participants were asked to provide self-explanations while predicting the next move in the prediction phase (Chi, de Leeuw, Chiu, & LaVancher, 1994; de Bruin, Rikers, & Schmidt, 2007; Renkl, 1997). Since the effect of self-explaining on learning to play chess is described elsewhere (de Bruin et al., 2007), the self-explanations were not further analysed.

After the prediction phase, participants proceeded to the judgement of learning phase (JOL phase, J1, J2, ... J6 in Table 1). This phase allowed learners to monitor their understanding of the previously studied endgame. Participants were shown all six positions (P1, P2, ... P6) from the previously studied endgame at which White was about to move. These positions were presented in random order, because ordering them correctly might stimulate learners to focus on the correct move that followed each JOL. That is, the correct move at J1 would have been revealed at J2, because the starting point for J2 is when move 1 just has been made by the computer. For every position, participants were asked to judge how well they had understood the move that had to be made: "How confident are you that you will predict a similar move correctly in the future?" (0% = definitely will not be able to predict, 20%, 40%, 60%, 80%, and 100% = definitely will be able to predict).

When participants had provided JOLs for all six moves (J1, J2, ... J6), they were shown the same chess exercise, and were once more asked to predict the moves White would make. This reprediction phase (R1, R2, ... R6) allowed participants to assess the accuracy of the JOLs, and enabled us to calculate metacomprehension accuracy by correlating individual JOLs with quality scores of repredictions. Because we used a mirror image position, the content and complexity of the chess exercise remained identical, but the appearance changed. Therefore, participants were forced to apply the correct principle, instead of reprediction based on memorisation from the prediction phase. Since participants were not intended to infer endgame principles in this phase, they did not self-explain and were given only 20 s to repredict each move.

In the move selection phase, which was designed to trigger self-regulation, participants were allowed to select moves for restudy. They were shown the positions of the mirror image exercise at which White was about to move one by one (S1, S2, ... S6), and pressed "y" if they wanted to study another example, and "n" if not. Participants were instructed to select at least two moves for restudy. Self-regulation was calculated by correlating individual JOLs with move selections (0 or 1). The restudy phase (RS1, RS2, ... RS6) consisted of studying the selected moves of another mirror image exercise, under the same instructions as the prediction phase. This procedure, starting with the prediction phase and ending with the restudy phase, was repeated for four chess exercises of equal difficulty. Participants worked through the application individually, which lasted on average 1 hour.

In the test phase that followed, participants were instructed to play four new KQK endgames against the computer and checkmate the black King in as few moves as possible. The first two were games of checkmate in six moves, whereas the final two were slightly more difficult (checkmate in eight and nine moves). Because the principles underlying the endgames were similar, the test exercises were analysed as a whole. We counted the number of times the participants were able to checkmate the black King. Participants received maximally 5 min per test exercise. Following the rules of the World Chess Federation (FIDE), a game ended in a draw when either 50 moves had been made and no piece was captured, the black King had taken the white Queen (checkmate impossible), stalemate had occurred, or the same position had appeared three times in the same game. Participants received maximally 5 min per test exercise.

Analysis

Participants' moves in the prediction, the reprediction, and the restudy phase of the exercises were compared to the computer's optimal move. When a predicted move was equal to or as good as the computer move (sometimes several good moves were possible) two points were awarded. Suboptimal moves that led to a game that required an extra move to checkmate were scored as one. For example, if a game could have been solved in six moves, but the move prediction was suboptimal and now seven moves were needed, one point was awarded. All other predictions were scored zero. The prediction scores are represented as percentages of the maximum score of 12 points per endgame of six moves.

Repeated measures analyses of variance were used to analyse the data in the learning phase. The dependent variables analysed were number of move selections, quality of (re)predictions, and number of checkmates in the test phase. Details of these analyses are provided in the Results section. Interaction effects were calculated, but will only be mentioned when significant. All alpha levels were set at .05.

The two main indicators of metacognitive activity in this experiment were metacomprehension accuracy and self-regulation. Since nearly half of the participants had at least one exercise with no variance in either JOLs or predictions (i.e., for a particular exercise the participant gave 40% JOLs for each move), metacomprehension accuracy was not calculated within chess exercises, but as an intraindividual Goodman-Kruskal gamma correlation between JOLs and repredictions across the four chess exercises for a total of 24 moves (Nelson, 1984). None of the participants had invariant JOLs, predictions, or move selections across the total of four chess exercises. Self-regulation was computed in a similar manner, as the intraindividual Goodman-Kruskal gamma correlations between JOLs and move selections across the four chess exercises. Independent samples t-tests were used to analyse differences between groups in metacomprehension accuracy, self-regulation, and number of checkmates in the test phase. Effect sizes for the ANOVAs were computed by means of η_p^2, for the t-tests by means of Cohen's d (Cohen, 1977).

RESULTS

Learning phase

As to the number of moves selected for restudy, a repeated measures analysis of variance with chess exercise (1, 2, 3, 4) as the within-subjects factor, and expertise level (inexperienced or experienced) as the between-subjects factor, showed a main effect of expertise level, $F(1, 36) = 4.84$, $MSE = 5.47$, $p < .05$, $\eta_p^2 = .12$. Inexperienced chess players selected more chess moves than experienced players. Moreover, an effect of chess exercise was found, $F(3, 36) = 16.38$, $MSE = 1.95$, $p < .001$, $\eta_p^2 = .31$. The number of move selections decreased over the four exercises.

To examine the accuracy of the move predictions, a repeated measures analysis of variance were performed on the prediction scores in the prediction and reprediction phase with chess exercise (1, 2, 3, 4) and prediction moment (prediction phase or reprediction phase) as within-subjects factors, and expertise level as the between-subjects factor. This analysis allowed us to assess whether participants' predictions improved across and within chess exercises. An overview of the prediction scores in the learning phase is provided in Table 2. A significant effect of group was

TABLE 2

Results of the learning phase of move selections and quality of move predictions for experienced (N = 20) and inexperienced (N = 18) chess players

Chess exercise	Mean move selections		Prediction score		Reprediction score		Restudy score	
	Inexperienced	Experienced	Inexperienced	Experienced	Inexperienced	Experienced	Inexperienced	Experienced
1	4.22 (1.11)	3.15 (1.53)	27.31 (15.34)	47.08 (17.58)	39.81 (17.28)	64.17 (20.25)	43.47 (19.60)	57.28 (25.44)
2	3.44 (1.92)	2.55 (0.83)	48.61 (13.78)	56.67 (13.94)	54.63 (21.81)	75.83 (18.32)	58.07 (26.15)	80.42 (20.51)
3	3.17 (1.82)	2.15 (1.31)	59.72 (14.92)	69.58 (17.79)	63.89 (17.62)	84.58 (11.56)	56.33 (20.84)	68.29 (34.33)
4	2.61 (1.91)	2.25 (1.45)	72.69 (16.12)	80.83 (17.95)	60.65 (18.26)	85.42 (15.74)	86.22 (21.34)	96.76 (9.54)
Total	3.36 (1.42)	2.59 (0.86)	52.08 (7.29)	63.54 (10.78)	54.75 (12.70)	77.50 (9.55)	56.77 (14.01)	75.60 (11.83)

Data are presented per chess exercise. Prediction score, reprediction score, and restudy score are presented as the percentage of correct predictions. Standard deviations in parentheses.

found, $F(1, 36) = 39.01$, $MSE = 568.51$, $p < .001$, $\eta_p^2 = .52$, which indicated that the experienced group had higher prediction scores than the inexperienced group. Also, an effect of chess exercise was observed, $F(3, 36) = 116.77$, $MSE = 333.74$, $p < .001$, $\eta_p^2 = .76$, showing that prediction scores improved over the chess exercises. An effect of prediction moment was observed, $F(1, 36) = 19.08$, $MSE = 274.37$, $p < .001$, $\eta_p^2 = .35$, which indicated that the reprediction scores were higher than the prediction scores. The interaction between prediction moment and condition was also significant, $F(3, 36) = 8.81$, $MSE = 274.67$, $p < .01$, $\eta_p^2 = .20$. This interaction indicated that the experienced group improved more from predictions to repredictions than the inexperienced group. The interaction between prediction moment and chess exercise was also significant, $F(3, 36) = 12.42$, $MSE = 261.61$, $p < .01$, $\eta_p^2 = .26$. This interaction showed that the difference between prediction and reprediction was large at the first exercise, but decreased over the following exercises.

Because the number of restudies differed between individuals, the restudy scores were analysed separately in a 4×2 (Chess exercise $\times$ Expertise level) repeated measures analysis of variance. A main effect of expertise was observed, $F(1, 28) = 8.44$, $MSE = 529.30$, $p < .01$, $\eta_p^2 = .23$, as well as a main effect of chess exercise, $F(3, 28) = 36.60$, $MSE = 496.86$, $p < .001$, $\eta_p^2 = .57$. However, since the latter analysis was based on the subset of participants who selected moves on all four exercises, results should be interpreted with caution (See Table 2).

Mean metacomprehension accuracy was .06 ($SD = .17$) for the inexperienced chess players and .23 ($SD = .26$) for the experienced chess players. To examine the extent to which these correlations indicated metacomprehension, we performed two one-sample t-tests to determine whether these correlations differed from zero. For inexperienced chess players, the metacomprehension correlation was not significantly different from zero, $t(17) = 0.06$, $p = .15$, $d = 0.36$. For experienced chess players, this correlation was significantly different from zero, $t(17) = 4.02$, $p < .01$, $d = 0.90$. Moreover, an independent samples t-test showed that the experienced chess players had higher metacomprehension accuracy than the inexperienced chess players, $t(36) = -2.35$, $p < .05$, $d = 0.79$. Mean self-regulation was $-.05$ for the inexperienced chess players ($SD = 0.21$) and $-.28$ for the experienced chess players ($SD = 0.20$). Again, this correlation was significantly different from zero for experienced chess players, $t(17) = -6.27$, $p < .001$, $d = 1.40$, but not for inexperienced chess players, $t(17) = -1.07$, $p = .30$, $d = 0.25$. The difference between experienced and inexperienced chess players was also significant, $t(36) = 3.41$, $p < .05$, $d = 1.61$.

Test phase

An independent samples t-test revealed that the experienced chess players checkmated the black King more often than the inexperienced chess players, $t(36) = -4.13$, $p < .001$, $d = 1.34$. On average, the experienced chess players checkmated the black King 2.95 times ($SD = 1.00$), and the inexperienced chess players 1.44 times ($SD = 1.25$).

DISCUSSION

The goal of the present study was to examine the effect of learner expertise on metacomprehension accuracy and quality of self-regulation. Looking at the learning gains first, we observed that the experienced chess players developed understanding of the endgame faster in the learning phase, and outperformed the inexperienced chess players in the test phase. With regard to the metacognitive measures, our data reveal that the experienced chess players showed higher metacomprehension accuracy and better self-regulation as indicated by the Goodman Kruskal gamma correlations. Moreover, the absolute level of metacomprehension accuracy and self-regulation of the inexperienced chess players was low: both measures were not significantly different from zero.

In sum, the accuracy of metacognitive activities does improve as a result of having proceduralised the rules, as Winne (1995) assumes. One possible explanation for this result is that the proceduralisation and subsequent automatisation frees up processing resources that can be allocated to monitoring understanding during learning. An alternative explanation could be that inexperienced players made more errors, and, therefore, were required to monitor more extensively. Given the higher working memory load these learners were faced with due to the absence of proceduralised subskills, the request to monitor understanding possibly interfered with learning. Of course, these two factors might have simultaneously exerted an influence. To further study the influence of limited processing resources in metacognitive processes, future research should include a measure of working memory load during learning, such as a dual task procedure that requires learners to perform a secondary task that increases memory load. The amount of interference that this task causes is indicative of the working memory load that is associated with the primary task (Brunken, Plass, & Leutner, 2003, 2004).

Having established a clear difference between inexperienced and experienced chess players in metacomprehension accuracy and self-regulation, we are left with one puzzling finding. The absolute level of these metacognitive processes in the experienced group, although significantly different from zero, was still relatively low. Especially when compared to levels of metacomprehension accuracy and self-regulation typically found in memorising word pairs or studying text (around .40, −.40, and higher; e.g., Thiede et al., 2003, 2005), the .23 and −.28 correlations observed in this study are considerably lower. A number of factors can be put forward to explain this. A first explanation is the relatively low level of experience our participants had with chess compared to participants in research on text comprehension or word pair learning. In the present study, the experienced group consisted of individuals who played chess on an irregular basis, sometimes not more than once a month. By contrast, participants in text comprehension research are usually university students who spend hundreds of hours studying texts each year. The metacognitive knowledge these participants have about reading texts is therefore probably more extensive than the knowledge the participants in the present study had about playing chess. The possible role of prior experience is substantiated by our finding that metacomprehension and self-regulation improved when learners had more experience with playing chess.

A second explanation could be that monitoring understanding and regulating study behaviour when acquiring a visual skill is more complicated than when acquiring a verbal skill as studying a text, or memorising word pairs. In the latter two situations, learners are required to make a retrieval attempt of the relevant information at the moment of judging comprehension. Attempting to retrieve the correct associated word or information from the text, activates cues that indicate what the chance is that the learner will accurately recall the information. According to prior research (Koriat, 1997; Thiede et al., 2003, 2005), metacomprehension accuracy depends on the activation of these cues that are needed when judging comprehension. These cues indicate, for instance, the ease with which the word pairs were processed (Shaw & Craik, 1989), or experienced familiarity of the items (Kelley & Lindsay, 1993). For these cues to have an effect, it is important that they are active at the time of judgement. This can be experimentally manipulated. For example, asking learners to generate keywords of a studied text after a certain delay, led to more accurate comprehension ratings and better self-regulation (Thiede et al., 2003). Thiede and colleagues explain the keyword effect in terms of access to the situation model of the text (Kintsch, 1988). The situation model provides an accurate predictor of long-term memory of the text, and, therefore, relying on information from the situation model improves metacomprehension accuracy. In more general terms, the keyword effect can be explained within the cue utilisation framework (Koriat, 1997) by posing that activating cues that are predictive of future performance improves metacomprehension. In contrast, when monitoring comprehension during chess skill acquisition, activation of cues relevant for judging comprehension is not achieved by making a retrieval attempt. Instead, learners have to try and mentally reinstate how easy or difficult it was to predict a certain move by for example trying to predict it again. That is, learners do not make a retrieval attempt, but rather a problem solving attempt. This problem solving attempt will activate cues about understanding that the learner needs to use to judge comprehension. The higher complexity of this monitoring process possibly explains part of its lower absolute accuracy. Self-regulation, being dependent on metacomprehension, will therefore also be poorer.

A third explanation that can be put forward indirectly stems from the enhanced complexity of the learning and monitoring situation, and concerns the design of the learning environment. In the present experiment, the judgement of learning phase was delayed until all moves had been studied. The same applied to the move selection (i.e., self-regulation) phase. This was mainly done because previous research revealed that relying on short-term memory for text or words is not an accurate predictor of long-term memory, and therefore, of test performance. Thus, when providing immediate JOLs, learners will base their judgements on short-term memory cues that are not predictive of long-term recall. However, since in cognitive skill acquisition the usual memory decay that is observed in text or word pair learning is not present, it might not be necessary to delay judgements of learning. Moreover, because judging comprehension when acquiring a cognitive skill as chess requires a problem solving attempt, delaying this judgement might lower the chance that learners will actually engage in this attempt. That is, because the initial prediction of the move does not directly precede the judgement moment, learners might not be inclined to perform a second problem solving attempt, and rely on superficial cues that are less predictive of performance when judging comprehension. To enhance chances of making a second problem solving attempt at time of judgement, the initial attempt should be made immediately prior to judgement. In the present study, this was not the case. The moment of judgement was delayed until all moves had been predicted, and the move selection phase was separated from the JOL phase by the reprediction phase. All in all, spacing the monitoring, self-regulation, and prediction phase, might have impeded learners to rely on relevant monitoring and self-regulation cues that stemmed from the prediction phase. To test this possibility, in future research the prediction and monitoring phase should be integrated, as well as the move selection phase. This will prevent the need for a second problem solving attempt and ensure that relevant monitoring cues are still active at the time of judgement.

In sum, the present study is the first to demonstrate an effect of expertise on metacomprehension accuracy and self-regulation within the cue utilisation framework (Koriat, 1997), and one of few to demonstrate an effect of individual differences on relative metacomprehension accuracy (for other examples, see Kelemen, Winningham, & Weaver, 2007 this issue; Miesner & Maki, 2007 this issue). As a result of prior domain knowledge, learners were better able to monitor comprehension and self-regulate study behaviour when learning to play a chess endgame, compared to learners who had no prior domain knowledge. However, we do not rule out the possibility that novices, under certain circumstances, are able to monitor their learning and regulate study behaviour, albeit to a limited extent. If the lack of sufficient processing resources is the main cause of the low metacomprehension accuracy and self-regulation, then adapting the learning environment to minimise processing requirements might improve metacognitive skills. For example, the abovementioned adjustment that leads to an integration of prediction and judgement phase might not only have an effect on experienced chess players, but also on novices. Future research is needed to provide an empirical test of this possibility.

In this study, the (meta)cognitive processes metacomprehension and self-regulation were operationalised in such a way that they could be statistically compared between groups. That is, unlike previous studies on the effect of expertise that analysed spontaneous metacognitive processes (e.g., Etelä-pelto, 1993; Prins et al., 2006; Schneider et al., 1998), learners were specifically stimulated to self-reflect and self-regulate their study behaviour. Based on knowledge of the processes underlying metacomprehension and self-regulation, an instruction was designed to stimulate use of these skills in learners. This study provides an example of how insights into cognitive mechanisms can be used to improve instruction in an educational setting. Future research should test the applicability of the cue utilisation framework in other skill domains besides chess, and, given the ultimate goal of transfer of learned skills to future situations, should also take into account long-term tests of performance.

REFERENCES

Anderson, J. R. (1983). *The architecture of cognition*. Cambridge, MA: Harvard University Press.

Azevedo, R., & Cromley, J. G. (2004). Does training on self-regulated learning facilitate students' learning with hypermedia? *Journal of Educational Psychology, 96*, 523–535.

Brunken, R., Plass, J. L., & Leutner, D. (2003). Direct measurement of cognitive load in multimedia learning. *The Educational Psychologist, 38*(1), 53–61.

Brunken, R., Plass, J. L., & Leutner, D. (2004). Assessment of cognitive load in multimedia learning with dual-task methodology: Auditory load and modality effects. *Instructional Science, 32*(1–2), 115–132.

Butler, D. L. (1998). The strategic content learning approach to promoting self-regulated learning. *Journal of Educational Psychology, 90*, 682–697.

Butler, D. L., & Winne, P. H. (1995). Feedback and self-regulated learning: A theoretical synthesis. *Review of Educational Research, 65*, 245–281.

Chi, M. T. H., de Leeuw, N., Chiu, M., & LaVancher, C. (1994). Eliciting self-explanations improves understanding. *Cognitive Science, 18*, 439–477.

Cohen, J. (1977). *Statistical power analysis for the behavioural sciences* (Rev. ed.). New York: Academic Press.

Cull, W. L., & Zechmeister, E. B. (1994). The learning ability paradox in adult metamemory research: Where are the metamemory differences between good and poor learners? *Memory and Cognition, 22*, 249–257.

De Bruin, A. B. H., Rikers, R. M. J. P., & Schmidt, H. G. (2005). Monitoring accuracy and self-regulation when learning to play a chess endgame. *Applied Cognitive Psychology, 19*, 167–181.

De Bruin, A. B. H., Rikers, R. M. J. P., & Schmidt, H. G. (2007). The effect of self-explanation and prediction on the development of principled understanding of chess in novices. *Contemporary Educational Psychology, 32*, 188–205.

Dunlosky, J., & Hertzog, C. (1997). Older and younger adults use a functionally identical algorithm to select items for restudy during multitrial learning. *Journal of Gerontology: Psychological Sciences and Social Sciences, 52*, 178–186.

Dunlosky, J., & Thiede, K. (2004). Causes and constraints of the shift-to-easier-materials effect in the control of study. *Memory and Cognition, 32*(5), 779–788.

Eteläpelto, A. (1993). Metacognition and the expertise of computer program comprehension. *Scandinavian Journal of Educational Research, 37*, 243–254.

Kelemen, W. L., Winningham, R. G., & Weaver, C. A., III. (2007). Repeated testing sessions and scholastic aptitude in college students' metacognitive accuracy. *European Journal of Cognitive Psychology, 19*, 689–717.

Kelley, C. M., & Lindsay, D. S. (1993). Remembering mistaken for knowing: Ease of retrieval as a basis for confidence in answers to general knowledge questions. *Journal of Memory and Language, 32*, 1–24.

Kintsch, W. (1988). The use of knowledge in discourse processing: A construction-integration model. *Psychological Review, 95*, 163–182.

Koriat, A. (1997). Monitoring one's own knowledge during study: A cue-utilization approach to judgments of learning. *Journal of Experimental Psychology: General, 126*, 349–370.

Koriat, A., Ma'ayan, H., & Nussinson, R. (2006). The intricate relationships between monitoring and control in metacognition: Lessons for the cause-and-effect relation between subjective experience and behavior. *Journal of Experimental Psychology: General, 135*(1), 36–69.

Maki, R. H. (1998). Test predictions over text material. In D. J. Hacker, J. Dunlosky, & A. C. Graesser (Eds.), *Metacognition in educational theory and practice* (pp. 117–144). Mahwah, NJ: Lawrence Erlbaum Associates, Inc.

Metcalfe, J., & Cornell, N. (2003). The dynamics of learning and allocation of study time to a region of proximal learning. *Journal of Experimental Psychology: General, 132*(4), 530–542.

Miesner, M. T., & Maki., R. H. (2007). The role of test anxiety on absolute and relative metacomprehension accuracy. *European Journal of Cognitive Psychology, 19*, 650–670.

Nelson, T. O. (1984). A comparison of current measures of the accuracy of feeling-of-knowing predictions. *Psychological Bulletin, 95*(1), 109–133.

Nelson, T. O., Dunlosky, J., Graf, A., & Narens, L. (1994). Utilization of metacognitive judgments in the allocation of study during multitrial learning. *Psychological Science, 5*, 207–213.

Prins, F. J., Veenman, M. V. J., & Elshout, J. (2006). The impact of intellectual ability and metacognition on learning: New support for the threshold of problematicity theory. *Learning and Instruction, 16*, 374–387.

Renkl, A. (1997). Learning from worked-out examples: A study on individual differences. *Cognitive Science, 21*, 1–29.

Schneider, W., Schlagmüller, M., & Vise, M. (1998). The impact of metamemory and domain-specific knowledge on memory performance. *European Journal of Psychology of Education, 13*(1), 91–103.

Shaw, R. J., & Craik, F. I. M. (1989). Age differences in predictions and performance on a cued recall task. *Psychology and Aging, 4*, 131–135.

Son, L. K., & Metcalfe, J. (2000). Metacognitive and control strategies in study-time allocation. *Journal of Experimental Psychology: Learning, Memory and Cognition, 26*(1), 204–221.

Thiede, K. W. (1999). The importance of monitoring and self-regulation during multitrial learning. *Psychonomic Bulletin and Review, 6,* 662–667.

Thiede, K. W., & Anderson, M. C. M. (2003). Summarizing can improve metacomprehension accuracy. *Contemporary Educational Psychology, 28,* 129–160.

Thiede, K. W., Anderson, M. C. M., & Therriault, D. (2003). Accuracy of metacognitive monitoring affects learning of texts. *Journal of Educational Psychology, 95,* 66–73.

Thiede, K. W., & Dunlosky, J. (1999). Toward a general model of self-regulated study: An analysis of selection of items for study and self-paced study time. *Journal of Experimental Psychology: Learning, Memory, and Cognition, 25*(4), 1024–1037.

Thiede, K. W., Dunlosky, J., Griffin, T. D., & Wiley, J. (2005). Understanding the delayed-keyword effect on metacomprehension accuracy. *Journal of Experimental Psychology: Learning, Memory and Cognition, 31*(6), 1267–1280.

Winne, P. H. (1995). Inherent details in self-regulated learning. *The Educational Psychologist, 30*(4), 173–187.

Winne, P. H. (2001). Self-regulated learning viewed from models of information processing. In B. Zimmerman & D. Schunk (Eds.), *Self-regulated learning and academic achievement: Theoretical perspectives* (2nd ed., pp.153–189). Mahwah, NJ: Lawrence Erlbaum Associates, Inc.

Zimmerman, B., & Schunk, D. (Eds.). (2001). *Self-regulated learning and academic achievement: Theoretical perspectives* (2nd ed.). Mahwah, NJ: Lawrence Erlbaum Associates, Inc.

EUROPEAN JOURNAL OF COGNITIVE PSYCHOLOGY
2007, 19 (4/5), 689–717

Repeated testing sessions and scholastic aptitude in college students' metacognitive accuracy

William L. Kelemen

Department of Psychology, California State University, Long Beach, CA, USA

Robert G. Winningham

Department of Psychology, Western Oregon University, Monmouth, OR, USA

Charles A. Weaver, III

Department of Psychology and Neuroscience, Baylor University, Waco, TX, USA

We performed three experiments to examine the effects of repeated study–judgement–test sessions on metacognitive monitoring, and to see if better students (those with higher Scholastic Aptitude Test or SAT scores) outperform low SAT students. In all experiments, mean metacognitive accuracy (bias scores and Gamma correlations) did improve with practice. Most improvement involved students' ability to predict which items would not be recalled later. In addition, students with high SAT scores recalled more items, were less overconfident, and adjusted their predictions more effectively. Thus, high SAT students may be able to adjust their metacognitive monitoring effectively without feedback, but low SAT students appear unlikely to do so. Educators may need to devise more explicit techniques to help low SAT students improve their metacognitive monitoring during the course of a semester.

The best college students effectively manage an assortment of requirements in a number of courses during a typical semester. Modes of instruction, course-related activities, and instructors' testing styles may vary from course to course even within a single major. As a result, many students greet the first test in a course with apprehension because they do not know exactly

Correspondence should be addressed to William L. Kelemen, Department of Psychology, 1250 Bellflower Blvd, Long Beach, CA 90840-0901, USA. E-mail: wkelemen@csulb.edu

Portions of Experiments 1 and 2 were presented at the 40th annual meeting of the Psychonomic Society in November 1999. The authors thank Rita Massey for retrieving students' SAT scores.

what to expect. Instructors may even counsel students who scored poorly on the first test that "You should do better next time, now that you know my testing style." Anecdotally, good students often seem better able to assess their preparation for their first exam, or at least to adjust to subsequent exams effectively. The present research was designed to test whether students' ability to predict their memory performance improves with repeated testing, and whether good students are indeed more effective at monitoring and adjusting their predictions compared with poor students.

Being able to predict performance accurately can impact test performance. For example, Thiede (1999) asked students to provide judgements of learning (JOLs) during study of vocabulary items followed by a recall test. Participants then selected items for restudy, and the process repeated for a total of five trials. Both monitoring accuracy and degree of self-regulation were significant predictors of test performance across trials. Similarly, Nelson, Dunlosky, Graf, and Narens (1994) demonstrated that recall was better when items receiving low JOLs were restudied compared with items receiving high JOLs. More recent work by Metcalfe and her colleague shows that good learners may identify a zone of "proximal learning", where additional study is most likely to produce the greatest gains (Metcalfe, 2002; Metcalfe & Kornell, 2003, 2005; Son & Metcalfe, 2000). Finally, Rawson and Dunlosky (2007 this issue) have developed a technique that can improve both monitoring and recall for definitions of key terms in text. Thus, the extant laboratory evidence suggests that one's degree of monitoring accuracy may influence academic success (for a field-based example, see Metcalfe, Kornell, & Son, 2007 this issue).

A separate, but related, question is whether or not good students (based on some index of learning ability) show superior monitoring accuracy compared with poor students. Such a finding should emerge if good students succeed at least in part because they have superior monitoring ability. Surprisingly, there is little evidence to support this idea. Lovelace (1984) found that JOL accuracy was unrelated to paired-associate recall. Cull and Zechmeister (1994) similarly found no differences in metacognitive strategies related to recall. Additional null results have been obtained using ease-of-learning judgements (Kearney & Zechmeister, 1989; Underwood, 1966), JOLs for texts (Pressley & Ghatala, 1990), and retrospective confidence judgements for general knowledge questions (Lichtenstein & Fischhoff, 1977). With few exceptions (e.g., Maki & Berry, 1984; Shaughnessy, 1979), differences in learning ability typically do not influence metacognitive accuracy.

Many past studies have assessed learning ability by computing a median split of participants based on the number of items recalled. This technique may be problematic because recall scores are used both to form the independent variable (learning ability group) and also to assess the dependent variable (monitoring accuracy). The present study used college students' scores on the Scholastic Aptitude Test (SAT, combined verbal and quantitative scores) to provide an independent, standardised estimate of academic ability.[1] Previous attempts to establish a relationship between SAT scores and metacognition have met limited success. Hall (2001) obtained nonsignificant correlations between SAT scores and responses to a metacognition questionnaire. When actual math or verbal problems involving metacognitive judgements were used, however, Tobias, Everson, and Laitusis (1999) reported correlations of 0.50 (math) and 0.29 (verbal) between metacognitive accuracy and SAT scores. In the Tobias et al. study, two metacognition questionnaires also were included but showed no significant correlations with SAT scores. Thus, questionnaires may be insufficient to detect this relationship, and so we obtained measures of actual metacognitive performance in the present study.

REPEATED TESTING AND METACOGNITION

Repeated testing in the classroom (e.g., administering weekly quizzes) has been shown to be an effective aid to course performance (e.g., Tuckman, 1996). In addition to classroom studies, there is a growing body of laboratory evidence showing the effectiveness of repeated testing on long-term memory accuracy (e.g., Butler & Roediger, 2007 this issue; Kang, McDermott, & Roediger, 2007 this issue; McDaniel, Anderson, Derbish, & Morrisette, 2007 this issue; Roediger & Karpicke, 2006). In regard to metacognitive monitoring, repeated study–test cycles of the same stimuli tend to produce higher levels of recall without commensurate increases in JOL magnitude, resulting in marked underconfidence (i.e., the "under-confidence-with-practice" effect, Koriat, Sheffer, & Ma'ayan, 2002). Most of the subsequent research (e.g., Koriat, Ma'ayan, Sheffer, & Bjork, 2006; Scheck & Nelson, 2005; Serra & Dunlosky, 2005) has continued to focus on repeated testing using the same stimulus materials. These studies are relevant to students restudying the same items for a single exam, but they do not speak to the question of how students use the outcome of one exam to prepare for subsequent ones on different material.

Some earlier studies on training examined repeated testing using different stimulus materials and showed increased metacognitive accuracy. Lichtenstein and Fischhoff (1980, Exp. 1) found that participants who completed 11 sessions (each containing 200 trials and followed by explicit feedback)

---

[1] Consistent with previous research (e.g., Frey & Detterman, 2004), we used SAT scores as a proxy for general intelligence test scores. The claim that SAT scores can be converted accurately to IQ scores is controversial, however (Bridgman, 2005).

showed increased calibration in their postdiction ratings, primarily in the early sessions. In Experiment 2, they obtained similar improvements using only three sessions of training and feedback. Zechmeister, Rusch, and Markell (1986) showed that just one session of training was enough to improve postdiction calibration. Furthermore, low-achieving students (as measured by classroom performance) initially were more overconfident than high-achieving students, and as a result, the low achievers benefited more from training.

Other research suggests that self-generated feedback alone can affect subsequent JOLs. Hertzog, Dixon, and Hultsch (1990) asked participants to predict future recall of categorised lists and texts. Two important findings emerged. First, the correlation between JOLs and recall increased across the three lists with practice alone ($rs = .24$, $.52$, and $.62$ for word recall, and $rs = .44$, $.54$, and $.58$ for texts). Second, structural equation modelling suggested that this improvement was due primarily to the effects of past recall performance. Koriat (1997) obtained an increase in item-by-item predictive accuracy using two different lists of paired associates. In Experiment 1, he found a small but reliable increase in mean G from List 1 ($G = .59$) to List 2 ($G = .65$). He attributed this increase to general experience with the task, but unfortunately no further experiments on practice were conducted using different items across trials. Because previous studies included only two or three practice trials completed during the same testing session, it is impossible to determine if further improvement would have been obtained using additional sessions, and if the improvements in JOL accuracy would extend to testing sessions completed at different times (as in a typical classroom situation).

GOALS AND HYPOTHESES

We examined the effects of repeated study–JOL–test sessions on metacognitive accuracy over a 5-week period. Participants completed five experimental sessions that involved learning foreign vocabulary items, making JOLs, and completing a recall test. Different vocabulary items were in each session, so these procedures were akin to the experiences of a student in a foreign language course having weekly vocabulary quizzes on new items each week. Two main issues were considered. First, we examined the effects of five study–JOL–test sessions on two aspects of predictive accuracy (i.e., estimates of overall recall and item-by-item JOL accuracy). Previous research suggests that monitoring accuracy may improve, but no studies have examined this issue directly, assessing both aspects of predictive accuracy over a period of weeks. Second, we examined the relationship between academic aptitude (SAT scores) and metacognitive monitoring. We predicted high SAT students would produce more accurate JOLs overall. If so, the benefits of practice might differ between groups, with low SAT students perhaps gaining more over time than high SAT students (cf. Zechmeister et al., 1986).

EXPERIMENT 1

Method

Participants and materials. Eighty-four undergraduates enrolled in Introductory Psychology volunteered to participate and received course credit. The stimuli were 100 pairs of Swahili–English translations (e.g., *WINGU–cloud*) drawn from Nelson and Dunlosky's (1994) norms. Five lists of 20 different vocabulary items were constructed to be of similar normative difficulty. The stimuli were projected on a large screen, and all participants were tested together in a lecture hall.

Design and procedure. Participants were informed that they would be learning foreign vocabulary and making judgements about their future memory performance. They also agreed to allow access to their SAT scores from their university records following completion of the study. Experimental sessions were held once per week, for 5 consecutive weeks. Before beginning each of the five sessions, participants were reminded that they would be asked to study 20 Swahili–English vocabulary items, provide a confidence rating about future recall for each item, and complete a memory test on the items. The to-be-learned vocabulary items appeared sequentially at a rate of 6 s/item. Immediately after studying an item, the Swahili cue word appeared below the following question, "How confident are you that in about 10 minutes you will be able to recall the English translation of the Swahili word below?" Participants circled one of the following six ratings on a separate judgement sheet: 0% confident (labelled "will not recall"), 20%, 40%, 60%, 80%, or 100% confident (labelled "will recall"). In order to discourage participants from rehearsing previous items during the study and rating phase, the Swahili cues did not appear on the participants' judgement sheets. Instead, the Swahili cues shown on the screen during the rating phase were numbered consecutively from 1 to 20, and participants circled their confidence rating listed next to the corresponding number. Participants were allowed 5 s to make each rating. A 2-s warning on the screen preceded presentation of the next vocabulary item. After studying and rating all 20 items, participants completed an unrelated filler activity for 5 min. Finally, a cued-recall test was distributed. The 20 Swahili cues were listed in a random order, and participants were asked to recall the English translation.

Participants were allowed as much time as necessary to complete the memory test.

Results and discussion

A total of 67 participants completed all five experimental sessions, and data from the remaining participants who did not complete all sessions were excluded. All tests of statistical significance were conducted at $p < .05$.

JOL magnitude.   Mean JOL ratings for each session were converted into proportions and appear in Table 1. Repeated measures analysis of variance (ANOVA) confirmed a significant decrease in JOL magnitude (i.e., confidence) over experimental sessions, $F(4, 264) = 17.69$, $MSE = 0.54$. A post hoc analysis was conducted to determine which sessions differed reliably. Differences between all pairs of means were compared to a critical difference, $q(4, 264) = 0.03$, computed using the Fisher-Hayter multiple-comparison test. Significant differences between sessions are noted in Table 1. Two findings emerged: (a) Participants showed higher mean confidence in Session 1 compared to all the other sessions, and (b) mean confidence in Session 2 was higher than in Sessions 4 and 5.

Cued recall performance.   Mean proportion correct on the memory test for each session was modest and consistent (means ranged from 0.31 to 0.35; see Table 1). No significant difference in recall across sessions was observed, $F(4, 264) = 1.36$, $MSE = 0.01$, suggesting that our attempt to equate the difficulty of stimuli across sessions was successful.

TABLE 1
Mean JOL magnitude, recall, and metacognitive accuracy across Sessions 1–5 in Experiment 1

Session	JOL magnitude*	Recall	Gamma*	Bias*
1	0.45_a	0.31	0.46_a	0.14_a
2	0.41_b	0.33	$0.58_{a, b}$	$0.08_{a, b}$
3	$0.38_{b, c}$	0.35	0.68_b	0.04_b
4	0.36_c	0.33	0.60_b	0.02_b
5	0.37_c	0.32	0.60_b	0.04_b

Main entries are mean values. Standard errors of the mean were less than 0.03 for all confidence, recall, and bias scores; standard errors of the mean were less than 0.05 for Gs. Entries sharing a common subscript within a column were not significantly different in post hoc analyses.
*A statistically significant effect was observed for that dependent variable, $p < .05$.

Metacognitive accuracy.   JOL accuracy can be assessed in several ways. Relative metacognitive accuracy is best measured by Goodman-Kruskal Gamma (*G*) correlations (Nelson, 1984). *G* is an ordinal measure of association that ranges from -1.0 to $+1.0$, with 0 indicating a complete lack predictive accuracy. *G* assesses item-by-item predictive accuracy, i.e., whether an item that received a high JOL is more likely to be recalled compared to a different item that received a lower JOL. Mean *G* levels are shown in Table 1.[2] A significant change was detected over sessions, $F(4, 256) = 4.47$, $MSE = 0.10$. A post hoc analysis compared all pairwise combinations of mean *G*s. Differences in mean *G* exceeding the critical difference, $q(4, 256) = 0.14$, are noted in Table 1. Relative metacognitive accuracy improved with practice: *G* was reliably higher in Sessions 3–5 compared to Session 1. This finding was somewhat surprising, because previous studies (e.g., Kelemen, Frost, & Weaver, 2000; Thompson & Mason, 1996) compared metacognitive accuracy over two sessions and found no increase in mean *G*. The present improvements in *G* emerged only when performance in Session 1 was compared to performance in Sessions 3–5. Thus, students may be able to improve their memory monitoring performance given enough practice in a particular task.

In addition to *G*, another way to assess metacognitive accuracy is to compare participants' mean confidence for all items to their mean recall. This procedure provides a measure of absolute metacognitive accuracy called bias.[3] A bias score greater than 0 indicates overconfidence, and a score less than 0 indicates underconfidence. Mean bias scores for all experimental sessions are listed in Table 1. All bias scores were positive, indicating general overconfidence in all sessions. ANOVA revealed a significant decrease in overconfidence over sessions, $F(4, 264) = 11.09$, $MSE = 1.41$. All pairwise combinations of means were compared to a critical difference, $q(4, 264) = 0.08$ using the Fisher-Hayter test. Participants were more overconfident in Session 1 than in Sessions 3–5. For example, by Session 5 the difference between participants' mean JOLs and mean recall was significantly reduced (0.04) compared to Session 1 (0.14). Thus, absolute metacognitive accuracy increased as participants became more familiar with the task.

Both measures of metacognitive accuracy presented so far (*G* and bias scores) are sensitive to changes in underlying levels of recall. Thus, it is

---

[2] Some *G*s could not be computed for two participants due to a lack of variability in recall. In Sessions 1 and 4, one participant failed to correctly recall any stimuli; in Session 5, one participant correctly recalled all 20 stimuli. Data from these participants were excluded from the following ANOVA and post hoc test.

[3] Bias scores are used frequently to assess absolute metacognitive accuracy, although statistical limitations of under- and overconfidence measures have been noted (Juslin, Winman, & Olsson, 2000).

possible that the differences in metacognitive accuracy over sessions could reflect changes in memory rather than memory monitoring. This possibility seems unlikely because we observed statistically significant changes in confidence across sessions but no significant changes in recall. Nevertheless, we conducted one further set of analyses to demonstrate that our findings reflected a genuine metacognitive phenomenon rather than a statistical artifact of recall. We examined the distributions of confidence ratings across the sessions, conditionalised on correct or incorrect recall (cf. Dougherty, Scheck, Nelson, & Narens, 2005). If monitoring accuracy improved across sessions, then the frequency distribution of JOLs should shift towards the ends of the scales (i.e., towards zero when subsequent recall was incorrect and towards 100 when subsequent recall was correct). Either type of shift would reflect a "pure" monitoring improvement because subsequent recall is held constant in each case (see Weaver & Kelemen, 1997, for a related discussion of these issues).

The top panel of Figure 1 confirms the predicted pattern: the percentage of incorrect items receiving JOLs of zero increased over Sessions 1–5. However, the curves for correct items (lower panel of Figure 1) were less conclusive. By and large, no systematic shifts emerged across sessions. Thus, the improvements in monitoring across sessions were real, and due primarily to participants' ability to recognise which items would not be recalled later.

SAT scores and performance.   In addition to changes in performance across sessions, we also examined whether high SAT students outperformed low SAT students. SAT scores were unavailable for seven participants; the mean for the remaining participants was 1110 ($SD = 150$). We compared students with low SATs (less than 1000, $n = 14$) to students with high SATs (greater than 1200, $n = 16$). These groups were roughly the bottom quartile ($M = 912$, $SD = 42$) and top quartile ($M = 1303$, $SD = 56$), respectively, of our sample. Summary statistics for these two groups are shown in Table 2. Overall, students with higher SATs tended to recall more on the memory test, and they were less overconfident.

Separate mixed-design ANOVAs were conducted on each of the four dependent measures. Session (1–5) was a within-subjects variable and group (low vs. high SAT) was included as a between-subjects variable. Because no significant interaction between session and group was detected in any of the ANOVAs, only main effects are discussed. For mean confidence, a significant decrease over sessions was obtained, $F(4, 112) = 8.62$, $MSE = 0.51$, but there was no significant difference in confidence between SAT groups. No significant main effects were obtained for recall, although Table 2 shows a clear trend toward higher recall in the high SAT group, $F(1, 28) = 3.23$, $MSE = 0.11$, $p = .08$. No significant main effects were

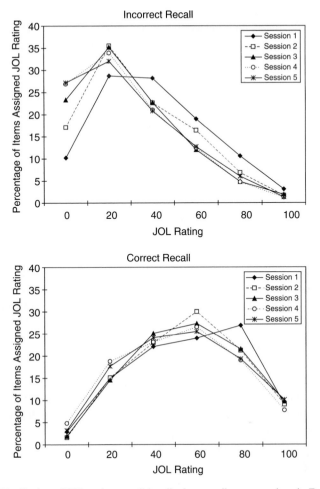

Figure 1. Distribution of JOL ratings conditionalised on recall across sessions in Experiment 1.

observed for G. Thus, no significant differences between high SAT students and low SAT students were detected in confidence, recall, and G.

Significant differences in bias were detected between groups and across sessions, $F(1, 28) = 7.76$, $MSE = 6.24$, and $F(4, 112) = 4.34$, $MSE = 1.25$, respectively. In both groups, overconfidence declined in later sessions, but low SAT students remained more overconfident than high SAT students all sessions. Even after four sessions of practice, low SAT students were no better (bias = 0.11) than unpractised high SAT students in Session 1 (bias = 0.09). In fact, bias scores were significantly greater than zero for low SAT students in Session 5 using a one-sample t-test, $t(13) = 2.81$, whereas high SAT students' bias scores did not differ significantly from zero in the final

TABLE 2
JOL magnitude, recall, and metacognitive accuracy as a function of SAT scores in
Experiment 1

Measure	Session				
	1	*2*	*3*	*4*	*5*
Confidence					
High SAT	0.44 (0.03)	0.41 (0.03)	0.37 (0.04)	0.35 (0.04)	0.33 (0.04)
Low SAT	0.45 (0.05)	0.42 (0.04)	0.38 (0.05)	0.36 (0.05)	0.38 (0.05)
Recall					
High SAT	0.36 (0.04)	0.38 (0.05)	0.36 (0.05)	0.39 (0.05)	0.40 (0.06)
Low SAT	0.29 (0.03)	0.29 (0.03)	0.29 (0.04)	0.28 (0.03)	0.28 (0.05)
Gamma					
High SAT	0.39 (0.08)	0.48 (0.08)	0.63 (0.10)	0.62 (0.09)	0.58 (0.10)
Low SAT	0.39 (0.13)	0.61 (0.09)	0.59 (0.09)	0.64 (0.07)	0.49 (0.14)
Bias*					
High SAT	0.08 (0.04)	0.03 (0.04)	0.02 (0.04)	−0.04 (0.03)	−0.07 (0.04)
Low SAT	0.16 (0.04)	0.14 (0.04)	0.09 (0.04)	0.09 (0.04)	0.11 (0.04)

Main entries are mean values; standard errors of the mean are in parentheses. In the "high SAT" group, SAT scores were greater than 1200; in the "low SAT" group, scores were less than 1000.
*A statistically significant difference between SAT groups was observed, $p < .05$.

session. These results contrast with Zechmeister et al.'s (1986) finding that poorer students benefited more from training; in our study, practice alone (in the absence of training) produced benefits only for the high SAT students.

In order to illustrate the differences in bias between groups, we plotted calibration curves using data from all five sessions (see Figure 2). These curves show actual recall as a function of predicted recall for both SAT groups. If participants were perfectly calibrated, they would recall none of the items given a $JOL = 0$, about 20% of items given a $JOL = 20$, and so forth. Examination of Figure 2 confirms that high SAT students were less overconfident than low SAT students across the entire rating scale.

EXPERIMENT 2

The first experiment showed significant improvement in memory monitoring accuracy over sessions, which we attributed to practice. However, an alternative explanation of these results is that, because the five lists of stimuli were comprised of different items, the range of item difficulty may have varied between sessions even though mean difficulty did not. Specifically, the variability could have been greater in Sessions 3–5

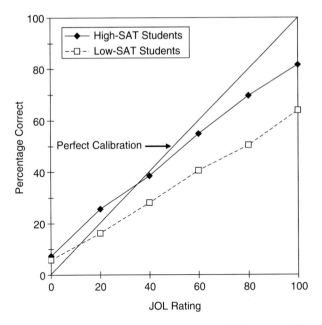

Figure 2. Mean recall as a function of predicted recall and SAT scores in Experiment 1.

compared with Session 1. For example, imagine Session 1 contained 20 items of roughly equal difficulty, whereas Session 5 contained 10 very easy items and 10 very difficult items. Mean recall might be the same for both sessions, but G likely would be higher in Session 5 because it contained two subgroups of items that could be discriminated from each other quite easily. This alternative does not explain why absolute metacognitive accuracy (bias scores) also improved over sessions. Nevertheless, Experiment 2 was designed to eliminate this potential confounding by varying the amount of practice between groups of participants, but comparing the same groups of items.

We formed three experimental conditions in Experiment 2. The procedures of Condition A were similar to those of Experiment 1: Participants studied lists of 20 Swahili–English pairs, made a JOL for each item immediately after study, and then received a cued-recall test. Participants completed five experimental sessions, each containing a unique set of 20 vocabulary items. In Condition B, participants also studied the lists of items and received memory tests, but they provided JOLs only for items in Session 5. The manipulation between Conditions A and B was whether participants made JOLs during Sessions 1–4. In Condition C, participants studied items, provided JOLs, and received a memory test, but they completed only Session 5. These three conditions are summarised in Table 3.

TABLE 3
Types of judgement provided by participants in each condition of Experiments 1–3

Experiment	Session				
	1	2	3	4	5
Experiment 1	item-by-item JOL	item-by-item JOL	item-by-item JOL	item-by-item JOL	item-by-item JOL
Experiment 2					
Condition A	item-by-item JOL aggregate JOL	item-by-item JOL aggregate JOL	item-by-item JOL aggregate JOL	item-by-item JOL aggregate JOL	item-by-item JOL aggregate JOL
Condition B	none	none	none	none	item-by-item JOL aggregate JOL
Condition C	N/A	N/A	N/A	N/A	item-by-item JOL aggregate JOL
Experiment 3					
Condition A	item-by-item JOL aggregate JOL	item-by-item JOL aggregate JOL	item-by-item JOL aggregate JOL	item-by-item JOL aggregate JOL	item-by-item JOL aggregate JOL
Condition B	item pleasantness aggregate JOL	item pleasantness aggregate JOL	item pleasantness aggregate JOL	item pleasantness aggregate JOL	item-by-item JOL aggregate JOL
Condition C	N/A	N/A	N/A	N/A	item-by-item JOL aggregate JOL

This design allowed us to test several hypotheses. First, we wanted to determine the source of memory monitoring improvement in Experiment 1: Did participants' monitoring improve because of practice, or was the increase an artifact of stimulus variability between sessions? If memory monitoring genuinely improves because of practice, then metacognitive accuracy in Session 5 should be reliably higher in Condition A compared with Condition C. The same set of items was used in each condition, but in Condition A participants had four sessions of practice, whereas participants in Condition C were naïve to the procedures.

If the improvement was genuine, then a second question concerns whether making JOLs is necessary, or whether merely being exposed to the stimuli, study procedures and cued-recall tests would improve memory monitoring. This issue can be addressed by comparing results from Conditions A and B in Session 5. In the former case, participants made JOLs during all five experimental sessions; in the latter, participants provided JOLs only in Session 5. If practice making JOLs is necessary, then metacognitive accuracy should be higher in Condition A.

We also obtained an additional type of prediction from our participants: a single "aggregate JOL" after studying all the items. Mazzoni and Nelson (1995) found that participants were less overconfident when making a single judgement about a collection of items than when their individual JOLs for each item were averaged. Perhaps the initial overconfidence in Experiment 1 reflected the averaging procedure we used, rather than a genuine metacognitive effect. We included both item-by-item and aggregate JOLs in Experiment 2 to test this possibility.

Method

Participants and materials.   A total of 135 undergraduates enrolled in Introductory Psychology volunteered for Experiment 2 (49 in Condition A, 47 in Condition B, and 39 in Condition C). All students received course credit for participation. The same five lists of Swahili–English items from Experiment 1 were used. Participants in each condition were tested as a group, but the three groups were tested separately.

Design and procedure.   Three experimental conditions were devised. In Condition A, participants completed one session per week, for 5 weeks. In each session, participants studied a list of 20 Swahili–English vocabulary items and provided a JOL immediately after studying each item. In order to increase the modest proportion of items recalled from Experiment 1, study time was increased to eight seconds for the Swahili–English pairs, and participants were allowed six seconds to provide each JOL. A 2-s warning preceded the presentation of each vocabulary item. After studying and rating all 20 items, participants were shown the following prompt, "Out of the 20 pairs you just studied, how many will you get right on a test in about 10 minutes?" This judgement was their aggregate JOL. Participants then completed a filler activity for 10 min and then completed a cued-recall test, as in Experiment 1.

Condition B was similar, except that participants did not provide JOLs during Sessions 1–4. Instead, the Swahili word alone appeared for six seconds below the prompt "Continue to think about the item below" immediately after study. The stimuli and all other procedures were the same as Condition A for Sessions 1–4. In Session 5, participants were informed that in addition to studying the 20 items, they would now be asked to predict their future memory performance. Thus, the stimuli and procedures in Session 5 were the same in Conditions A and B.

Only one experimental session was included in Condition C. Participants were instructed to study the items and make judgements about their future memory performance. The stimuli and procedures were the same as those used in Session 5 of previous conditions.

Results and discussion

A total of 35 participants completed all five sessions in Condition A, 28 completed all five sessions in Condition B, and 39 participants completed Condition C (which included only one session). We included all available data from participants in the following statistical analyses, which resulted in small fluctuations in sample sizes across sessions for Conditions A and B.

Changes in JOLs over sessions.   Participants in Condition A provided JOLs during each of the five sessions. Two types of JOLs were elicited from participants: (a) a judgement for each vocabulary item (item-by-item JOLs), and (b) one judgement about how many of the 20 words would likely by recalled (an aggregate JOL). One participant failed to provide an aggregate JOL in Session 2, and another participant failed to provide one in Session 4. Both types of JOL were converted into proportions to permit direct comparison.

We first considered item-by-item JOLs. Data from 35 participants in Condition A who completed all five sessions were analysed. Confidence appeared to decrease after the initial session: mean item-by-item JOLs were 0.44, 0.37, 0.38, 0.35, and 0.37, for Sessions 1–5, respectively (the standard error of the mean, SEM, for each value was less than 0.03). Consistent with Experiment 1, a repeated measures ANOVA revealed a significant change, $F(4, 136) = 10.65$, $MSE = 0.01$. A critical difference between means was calculated using the Fisher-Hayter post hoc test, $q(4, 136) = 0.04$. This post hoc test confirmed that confidence was higher in Session 1 than in all subsequent sessions.

A similar pattern of results emerged using aggregate JOLs. Confidence again decreased after the first session (mean aggregate JOLs for Sessions 1–5 = 0.41, 0.35, 0.38, 0.36, 0.35; all $SEMs < 0.03$). A repeated measures ANOVA on data from 33 participants who provided aggregate JOLs for all five sessions showed a significant change, $F(4, 128) = 3.51$, $MSE = 0.01$. A critical difference between means was computed, $q(4, 128) = 0.05$. Aggregate JOLs in Session 1 were reliably higher compared to those in Sessions 2, 4, and 5. In short, the magnitude of both mean item-by-item and aggregate JOLs decreased after the initial session.

JOL magnitude between conditions. The first two columns of Table 4 contain mean JOLs from Session 5 between conditions. These values represent the effects of practice on confidence. For item-by-item JOLs, a one-way between-groups ANOVA was conducted using 104 participants who provided judgements for all items from Session 5. A significant effect of condition was obtained, $F(2, 101) = 8.73$, $MSE = 0.03$. Participants with previous JOL experience (Condition A) were less confident than participants with no experience in making JOLs (Conditions B and C), $q(2, 101) = 0.05$.

A similar pattern emerged for aggregate JOLs. A significant effect of condition was found, $F(2, 100) = 9.36$, $MSE = 0.03$, and a critical difference between means was computed, $q(2, 100) = 0.05$. The magnitude of aggregate JOLs differed reliably across all three conditions. Participants with JOL experience were the least confident, and participants in Condition B were the most confident.

Cued recall performance. Participants in Conditions A and B learned vocabulary items and completed cued-recall tests during Sessions 1–5. Mean recall in Condition A was 0.34, 0.41, 0.43, 0.39, and 0.42 for Sessions 1–5, respectively. Corresponding recall values for Condition B were 0.42, 0.52, 0.50, 0.53, and 0.42 ($SEMs$ for both groups were all less than 0.03). A 2×5 (condition by session) ANOVA was conducted. Recall was reliably higher in Condition B compared to Condition A, $F(1, 60) = 7.84$, $MSE = 0.12$. This may reflect a slight difference in procedures between conditions during Sessions 1–4. Participants in Condition A studied the cue–target pair for 8 s and then made a cue-alone JOL for 6 s; participants in Condition B studied the cue–target pair for 8 s and then continued to study the cue alone for 6 s. The extra study time in Condition B may have increased cued-recall performance. A significant effect of session was noted, $F(4, 240) = 4.99$, $MSE = 0.01$, and there was a significant interaction between condition and session, $F(4, 240) = 3.48$, $MSE = 0.01$.

TABLE 4
JOL magnitude, recall, and metacognitive accuracy for Session 5 in Experiment 2 by condition

Task experience	Item-by-item JOLs*	Aggregate JOLs*	Recall	Gamma*	Bias (item-by-item)*	Bias (aggregate)*
Condition A (four sessions of JOLs and recall)	0.37 (0.02)$_a$	0.35 (0.02)$_a$	0.42 (0.03)	0.51 (0.08)$_a$	−0.05 (0.03)$_a$	−0.06 (0.03)$_a$
Condition B (four sessions of recall only)	0.50 (0.03)$_b$	0.54 (0.04)$_b$	0.42 (0.04)	0.52 (0.08)$_a$	0.08 (0.03)$_b$	0.10 (0.03)$_b$
Condition C (no previous experience)	0.52 (0.03)$_b$	0.44 (0.03)$_c$	0.35 (0.03)	0.26 (0.06)$_b$	0.17 (0.05)$_c$	0.09 (0.03)$_b$

Main entries are mean values; standard errors of the mean are in parentheses. Entries sharing a common subscript within a column were not reliably different in post hoc analyses.
*A statistically significant effect was observed, $p < .05$.

More important were potential differences in recall between conditions in Session 5 (see third column in Table 4). In the final session, all three conditions received 8 s of study time and 6 s to make a JOL. ANOVA showed no significant differences in recall between conditions, although there was a trend for practised participants to recall more than unpractised participants. Thus, when study time between groups was equal (Session 5), no differences in recall were observed.

Relative metacognitive accuracy (G). Gamma correlations were calculated between JOLs and recall for 35 participants in Condition A. Mean values across sessions were 0.53, 0.61, 0.65, 0.60, 0.51 (*SEMs* ranged from 0.04 to 0.08). The relatively low metacognitive accuracy in Session 5 failed to replicate Experiment 1 and in contrast to those findings, no significant change in metacognitive accuracy was found across sessions, $F(4, 136) = 1.07$, $MSE = 0.11$.

The analysis of major interest, however, was the comparison of *G*s across Conditions A–C in Session 5. The mean values are listed in Table 4. A significant difference in *G* between conditions was detected, $F(2, 99) = 4.26$, $MSE = 0.18$. A critical difference was calculated post hoc, $q(2, 99) = 0.12$, which confirmed that mean *G* was higher in Conditions A and B (*G*s = 0.51 and 0.52, respectively) than in Condition C ($G = 0.26$). These results suggest that practice with the experimental procedures, but not necessarily practice making JOLs, underlies the increase in relative memory monitoring accuracy over sessions. Further, these results are inconsistent with the notion that stimulus artifacts caused the increase across sessions in Experiment 1, because in this case significant differences emerged between conditions using the same stimuli in each group.

Absolute metacognitive accuracy. Bias scores (i.e., the signed difference between confidence and accuracy) were computed for each participant in Condition A using two measures of confidence: (a) the mean of item-by-item JOLs, and (b) aggregate JOLs. Mean bias scores for Sessions 1–5 using item-by-item JOLs were 0.10, −0.03, −0.05, −0.03, and −0.05, respectively. Using aggregate JOLs, mean bias scores were 0.06, −0.06, −0.05. −0.02, and −0.06 (all *SEMs* were less than 0.03). ANOVA revealed a significant effect of session on bias, $F(4, 136) = 11.96$, $MSE = 0.02$ for item-by-item JOLs, and $F(4, 128) = 8.69$, $MSE = 0.02$ for aggregate JOLs. A critical difference between means of 0.08 was calculated for both measures of bias. These post hoc tests indicated a significant difference between Session 1 and Sessions 2–5. As in Experiment 1, participants were most overconfident in Session 1.

We also examined differences in bias between conditions in Session 5 (see Table 4). ANOVA revealed a significant difference between conditions, $F(2, 101) = 7.56$, $MSE = 0.06$ for item-by-item JOLs, and $F(2, 101) = 9.82$,

$MSE = 0.03$ for aggregate JOLs. Critical differences between the means were calculated, $q(2, 101) = 0.06$ for item-by-item JOLs and $q(2, 101) = 0.05$ for aggregate JOLs. Significant differences between means are noted in Table 4. In general, participants with JOL experience (Condition A) were slightly underconfident, compared to the overconfidence demonstrated by participants without JOL experience (Conditions B and C). This pattern was consistent whether bias was measured using aggregate JOLs or mean item-by-item JOLs.

These findings suggest that the reduction in bias scores across sessions in Experiment 1 did not reflect stimulus artifacts. Further, this pattern was consistent using both item-by-item JOLs as well as a single rating based on the aggregate of items. In contrast to the post hoc tests on G scores, however, experience with the study and recall procedures alone (Condition B) did not improve performance. In fact, participants were most overconfident in Condition B. This finding may be related to the fact that participants in Condition B were instructed to continue studying each item (rather than make a JOL) during Sessions 1–4. These procedures produced higher levels of recall in the first four sessions compared to Condition A, and may have produced overconfidence in Session 5. One way to address this issue would be to elicit JOLs in Condition A and stimulus ratings on some unrelated dimension (e.g., pleasantness) in Condition B. We followed these procedures in Experiment 3 to control this potential confounding.

As a final measure of metacognitive accuracy, we again plotted the distributions of confidence ratings conditionalised on correct or incorrect recall (see Figure 3). The top panel shows the distribution of rating for unrecalled items: Participants in Condition A were substantially more likely to assign these items low JOLs. This pattern supports our other analyses and suggests better metacognitive monitoring for unrecalled items in Condition A. For items that were recalled successfully (bottom panel of Figure 3), however, participants in Condition A remained the least confident. Thus, the major effect of practice making JOLs on absolute measures of metacognition was to reduce confidence overall. These changes in confidence also support our interpretation that the reduction in bias scores was due to changes in metacognitive judgements and not the levels of recall.

SAT scores and performance.   The mean SAT score for participants was 1101 $(SD = 133)$.[4] We compared the performance of low SAT students (SAT $= 1010$ or below; $n = 33$) to high SAT students (SAT $= 1200$ or above;

---

[4] SAT scores were unavailable for 19 of the 135 participants. For 12 of these students, scores from the American College Test (ACT) were obtained and converted to equivalent SAT scores using a conversion table provided by the Institutional Research and Testing department at Baylor University. Data from the remaining seven students were omitted.

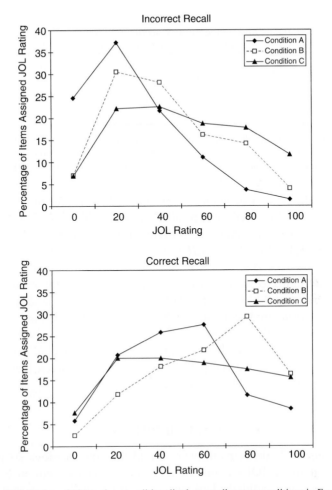

Figure 3.   Distribution of JOL ratings conditionalised on recall across conditions in Experiment 2.

$n = 31$). These two groups corresponded to approximately the bottom and top quartiles of our entire sample ($N = 128$), respectively. The mean score for low SAT students was 940 ($SD = 70$) and the mean for high SAT students was 1271 ($SD = 76$).

Twenty-four participants from each SAT group completed all sessions, including Session 5. Mean confidence, recall, and metacognitive accuracy by group are shown in Table 5. ANOVAs were performed on these data to test for differences in performance between high and low SAT students.[5]

---

[5] Before computing the ANOVAs, a chi-square test for independence was performed on condition and SAT group. No reliable effect was observed, $\chi^2(2) = 3.69$, $p > .15$, so ANOVAs were computed using data summed across all three conditions.

TABLE 5
JOL magnitude, recall, and metacognitive accuracy in Session 5 of Experiment 2 for high versus low SAT students

Group	Item-by-item JOLs	Agg-JOL	Recall*	Gamma	Bias (item-by-item)*	Bias (aggregate)*
High SAT	0.45 (0.03)	0.45 (0.03)	0.48 (0.04)	0.30 (0.10)	−0.03 (0.03)	−0.03 (0.03)
Low SAT	0.47 (0.04)	0.42 (0.04)	0.28 (0.03)	0.41 (0.09)	0.19 (0.05)	0.14 (0.04)

Main entries are mean values; standard errors of the mean are in parentheses. In the "high SAT" group, SAT scores were 1200 or greater; in the "low SAT" group, scores were 1010 or less. *A statistically significant difference between SAT groups was observed, $p < .05$.

Significant differences between groups emerged for recall (high SAT mean recall = 0.48 and low SAT mean recall = 0.28), $F(1, 46) = 17.07$, $MSE = 0.03$. High SAT students also showed less overconfidence as measured by item-by-item JOLs, $F(1, 46) = 9.42$, $MSE = 0.07$ and aggregate JOLs, $F(1, 46) = 11.33$, $MSE = 0.03$. Bias scores in Session 5 were substantially larger and significantly nonzero for low SAT students, bias = 0.19 for mean item-by-item JOL bias and 0.14 for aggregate JOL bias, $t(23) = 3.81$ and $t(23) = 3.53$, respectively, compared to high SAT students who showed bias scores that were not significantly different from zero, bias = −0.03 for both measures. No significant differences between groups emerged for confidence or Gs. These findings replicated the pattern of differences between high and low SAT groups obtained in Experiment 1.

EXPERIMENT 3

In Experiment 2, participants who completed four sessions of practice, but did not make JOLs (Condition B), were overconfident in Session 5 when JOLs were required. This may have occurred because participants were allowed to study items for longer in Sessions 1–4 than in Session 5. The purpose of Experiment 3 was to control for this potential confound. We formed three experimental conditions in Experiment 3. In Condition A, participants completed five sessions of study–JOL–test procedures as before. In Condition B, participants studied each item and provided a pleasantness rating (instead of a JOL), followed by a test, during Sessions 1–4. During Session 5, participants followed the same study–JOL–test procedures as Condition A. Finally, participants in Condition C completed only Session 5, as before. These procedures are summarised in Table 3.

Method

Participants and materials. A total of 118 college undergraduates volunteered for Experiment 3 (54 in Condition A, 40 in Condition B, and 24 in Condition C). All students received course credit for participation. The same 100 Swahili–English items from previous experiments were used, but five new lists were created. Participants in each condition were tested as a group, but the three groups were tested separately.

Design and procedure.   Three experimental conditions were devised. The procedures for Condition A were identical to Experiment 2. In Condition B, participants provided ratings of item pleasantness, ranging from 1 ("very unpleasant") to 6 ("very pleasant"), during Sessions 1–4. Participants were instructed to provide JOLs instead of pleasantness ratings in Session 5. Thus, the stimuli and procedures during Session 5 were the same in Conditions A and B. In Condition C, participants completed only Session 5 to provide an unpractised comparison group. To maintain consistency across sessions and conditions, participants always provided aggregate JOLs. Participants in Conditions A and B were tested twice per week (on Monday and Friday), because testing did not begin until late in the semester. SAT scores were not obtained.

Results

A total of 40 participants in Condition A and 21 participants in Condition B completed all five sessions. Twenty-four participants completed Condition C, which included only one session. Five participants reported participating in a previous version of this experiment on a posttest questionnaire during Session 5. Data from these participants and all others who did not complete Session 5 were omitted from statistical analyses. As a result, the final sample sizes for Conditions A–C were 36, 20, and 24, respectively.

JOL magnitude and recall.   Confidence was assessed using mean item-by-item JOLs and aggregate JOLs in Session 5 (see Table 6). One participant did not provide an aggregate JOL during Session 5. Although there was a trend towards higher JOL magnitude for participants in Condition C, no significant differences were observed between groups using either measure, $F(2, 77) = 1.62$, $MSE = 0.02$, for item-by-item JOLs and $F(2, 76) = 1.23$, $MSE = 0.03$, for aggregate JOLs. No significant differences in recall were found between groups, $F(2, 77) = 0.82$, $MSE = 0.04$.

Metacognitive accuracy.   As in Experiments 1 and 2, two measures of metacognitive accuracy (G and bias) were calculated. Mean values for each measure are reported in Table 6. A significant effect of condition on G was obtained, $F(2, 77) = 4.73$, $MSE = 0.10$. Post hoc tests revealed that participants who received practice (Conditions A and B) showed higher G than unpractised participants (Condition C), $q(2, 77) = 0.10$. This finding was consistent with previous results showing an improvement due to practice with the study and test procedures. Making JOLs during Sessions 1–4 (versus pleasantness ratings) did not improve G during Session 5. In fact, G was higher in Condition B than Condition A in the final session.

The main purpose of Experiment 3 was to control for an increase in study time in Condition B of Experiment 2 that may have produced over-confidence (bias). Holding study time constant in Experiment 3, no differences in bias scores were found between Conditions A and B (see Table 6). Unpractised participants in Condition C were more overconfident than participants in the other conditions. This pattern is replicated in the top panel of Figure 4, which shows that participants in Conditions A and B were more likely to use low JOLs for items that were not subsequently recalled. For mean item-by-item bias, a significant effect of condition emerged, $F(2, 77) = 3.27$, $MSE = 0.04$. A significant difference was also found using aggregate JOLs, $F(2, 76) = 4.64$, $MSE = 0.03$. Post hoc tests showed that for both measures of bias, unpractised participants were more overconfident than practised participants, $q(2, 77) = 0.06$ and $q(2, 76) = 0.06$. These results suggest that the overconfidence in Experiment 2 was indeed due to the difference in study time. Experiment 3 also confirmed previous main findings that unpractised participants (Condition C) were most over-confident and showed the lowest Gs. Finally, the distributions of JOLs conditionalised on correct or incorrect recall again showed that most of the systematic changes in metacognitive judgements occurred for items that were not correctly recalled subsequently.

GENERAL DISCUSSION

We examined the effects of multiple study–JOL–test sessions on students' metacognitive performance for foreign vocabulary items over a 5-week period. Participants completed five experimental sessions, and different vocabulary items were included in each session. In all three experiments, the predictive accuracy of JOLs increased with practice. Improvements in memory monitoring were observed using mean G scores, which assessed item-by-item JOL accuracy, as well as bias scores, which assessed global under- and overconfidence. Most of the increase in metacognitive accuracy occurred in the first three sessions, but performance remained high in

TABLE 6
JOL magnitude, recall, and metacognitive accuracy in Session 5 of Experiment 3 as a function of previous experience

Task experience	Item-by-item JOLs	Aggregate JOLs	Recall	Gamma*	Bias (item-by-item)*	Bias (aggregate)*
Condition A (four sessions of JOLs and recall)	0.42 (0.02)	0.40 (0.03)	0.40 (0.03)	$0.53 (0.06)_a$	$0.02 (0.03)_a$	$0.0 (0.02)_a$
Condition B (four sessions of ratings and recall)	0.44 (0.04)	0.40 (0.04)	0.39 (0.05)	$0.65 (0.06)_b$	$0.05 (0.04)_a$	$-0.01 (0.03)_a$
Condition C (no previous experience)	0.49 (0.03)	0.46 (0.04)	0.34 (0.04)	$0.36 (0.06)_c$	$0.15 (0.05)_b$	$0.13 (0.05)_b$

Main entries are mean values; standard errors of the mean are in parentheses. Entries sharing a common subscript within a column were not reliably different in post hoc analyses.
* A statistically reliable effect was observed, $p < .05$.

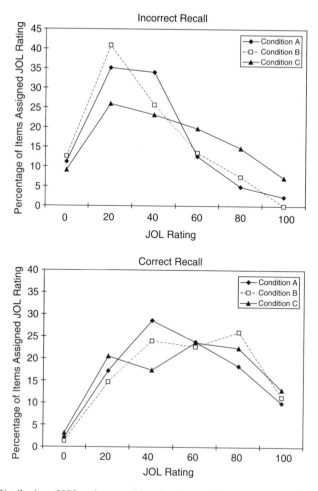

Figure 4. Distribution of JOL ratings conditionalised on recall across conditions in Experiment 3.

subsequent sessions. In contrast to past research (e.g., Zechmeister et al., 1986), these increases were achieved without any special training or explicit feedback.

Experiments 2 and 3 were designed to determine whether making JOLs during practice sessions was necessary to increase monitoring accuracy. In both experiments, practice making JOLs was not required to increase mean G. For bias, overconfidence was observed in Experiment 2 when participants did not have previous JOL experience, suggesting that JOL practice may affect absolute monitoring accuracy. When participants were asked to provide pleasantness ratings during the first four sessions of Experiment 3, however, bias in Session 5 was no greater than when participants made JOLs during Sessions 1–4. Thus, exposure to the study and recall procedures alone was sufficient to improve both *G* and bias scores; making JOLs during previous sessions did not increase monitoring accuracy further.

Participants tended to be overconfident during the initial testing session in all three experiments, so their adjustments in absolute metacognitive accuracy required a reduction in JOL magnitude in subsequent sessions. Plots of JOLs conditionalised upon recall confirmed that practice improved participants' ability to identify items not likely to be remembered and to assign these items low JOLs. In contrast, practice did not systematically increase JOL accuracy for correctly recalled items. It is simplistic, though probably accurate, to say that practice helped participants determine which items they would not remember on subsequent tests. These adjustments in calibration were impressive nevertheless: participants were able to estimate how many items they had answered correctly on a test (e.g., 7 out of 20, or 35%), and then adjust (a) their subsequent aggregate JOL (e.g., predict 7 out of 20 next week), and (b) the magnitude of all 20 JOLs a week later such that the mean rating was appropriate (e.g., the mean of all JOLs next week would be about 35%). This latter adjustment seems especially remarkable, especially given that participants received no feedback or instructions to adjust their future judgements.

Koriat (1997) has proposed that participants may use one or more classes of cue when making JOLs. In the present research, JOLs in the initial session may have been based on intrinsic cues (e.g., perceived relatedness between the items, etc.), but in later sessions extrinsic cues (e.g., conditions of learning including study time, retention interval, etc.) might have become available to enhance predictive accuracy. Experience with the recall test may have provided especially important information for participants. Vye, Schwartz, Bransford, Barron, and Zech (1998) reported on a programme to increase students' metacognitive awareness and argued that knowing what to expect on a test provides critical cues for future study. They claimed, "The better individuals can imagine (model) the situation in which they must use their knowledge, the easier it is for them to assess their level of preparation. *Increasing experience with particular situations* ... increases the accuracy with which one can anticipate the kinds of knowledge and skills necessary to perform adequately" (p. 309, italics added). A similar mnemonic debiasing account of practice on JOLs has been proposed by Koriat et al. (2006). In future research, it might be possible to tease apart the relative contributions of knowledge about study versus test procedures by comparing groups with practice in each of these components. Regardless of the precise theoretical mechanism, the metacognitive benefits of administering multiple realistic practice tests (or quizzes) to students in the classroom can be inferred from these laboratory data.

Learning ability and metacognition

A final aim of the present study was to determine whether individual differences in confidence, recall, and metacognition were related to learning ability (as indexed by SAT scores). No differences in confidence were detected between low and high SAT students, but recall did differ according to SAT scores. In other words, low SAT students remembered less than their high SAT counterparts, but they were just as confident in their future memory performance, resulting in marked overconfidence in the former group. This relative overconfidence was observed across the rating scale (see Figure 2). In addition, low SAT students failed to adequately reduce their overconfidence in later sessions. Even during the final session, low SAT students were more overconfident than unpractised high SAT students. Future research is required to determine the sources of these persistent errors. Perhaps low SAT students were less adept at constructing self-generated feedback, or perhaps they were less sensitive to their own feedback concerning previous recall.

High SAT students, conversely, did reduce their overconfidence across sessions, and even trended toward underconfidence by Session 5 in both experiments. This pattern is analogous to the "underconfidence-with-practice" effect for repeated stimuli reported by Koriat et al. (2002), which has been the subject of scrutiny (e.g., Koriat et al., 2006; Scheck & Nelson, 2005; Serra & Dunlosky, 2005). Before now, this effect was not thought to extend to repeated testing using different items (Koriat et al., 2002). One possible explanation for our results relates to Scheck and Nelson's (2005) observation that underconfidence with practice was less likely to occur when levels of recall were very low and more likely to occur when recall was high. Scheck and Nelson proposed that anchoring effects for JOLs contributed to the underconfidence-with-practice effect. Consistent with their account, underconfidence with practice emerged only for our high SAT students, who showed higher levels of recall but equivalent JOL magnitude compared with the rest of the sample. The fact that underconfidence with practice emerged only in our participants with the highest levels of recall suggests that anchoring effects may be at work; other proposed accounts of this phenomenon would require modification to account for the effects of different items on each trial.

No consistent relationship was found between G and SAT scores, consistent with previous studies comparing learning ability and item-by-item monitoring accuracy (e.g., Lovelace, 1984). However, the reliability of G in testing for individual differences has been found lacking in past research (e.g., Kelemen et al., 2000; see also Higham & Arnold, 2007 this issue, for a related finding). It may be unrealistic to expect G to correlate with learning ability, or any other psychological variable, until a procedure for obtaining reliable differences is devised.

Educators interested in improving student learning are well acquainted with the so-called "Matthew effect": Interventions designed to improve the performance of poor students provide an even greater benefit to good students (e.g., Stanovich, 1986). Here we find yet another example: Good students do better on tests, and are also better predictors of their future performance. One intervention designed to improve predictions—repeated practice—benefited those students who were already better at predicting their future performance to begin with. Low SAT students performed worse in initial memory tests, and failed to benefit from additional practice.

Many current theories of metamemory hypothesise that one source of JOLs is implicit retrieval of the target item. Failure to retrieve an item at time of JOL is highly diagnostic of future memory failure (see Weaver, Terrell, Krug, & Kelemen, in press). Our results suggest that one way in which low SAT students differ is their inability to benefit from such retrieval failures. Although these students may have the impression that they will do better after they have learned an instructor's "testing style", our findings suggest that this, too, is an illusion. Therefore, instructors may need to provide explicit feedback or metacognitive training to low SAT students for them to benefit from repeated testing.

REFERENCES

Bridgman, B. (2005). Unbelievable results when predicting IQ from SAT scores: A comment on Frey and Detterman (2004). *Psychological Science, 16*, 745–746.

Butler, A. C., & Roediger, H. L., III. (2007). Testing improves long-term retention in a simulated classroom setting. *European Journal of Cognitive Psychology, 19*, 514–527.

Cull, W. L., & Zechmeister, E. B. (1994). The learning ability paradox in adult metamemory research: Where are the metamemory differences between good and poor learners? *Memory and Cognition, 22*, 249–257.

Dougherty, M. R., Scheck, P., Nelson, T. O., & Narens, L. (2005). Using the past to predict the future. *Memory and Cognition, 33*, 1096–1115.

Frey, M. C., & Detterman, D. K. (2004). Scholastic assessment or g? The relationship between the Scholastic Assessment Test and general cognitive ability. *Psychological Science, 15*, 373–378.

Hall, C. W. (2001). A measure of executive processing skills in college students. *College Student Journal, 35*, 442–450.

Hertzog, C., Dixon, R. A., & Hultsch, D. F. (1990). Relationships between metamemory, memory predictions, and memory task performance in adults. *Psychology and Aging, 5*, 215–227.

Higham, P. A., & Arnold, M. M. (2007). How many questions should I answer? Using bias profiles to estimate optimal bias and maximum score on formula-scored tests. *European Journal of Cognitive Psychology, 19*, 718–742.

Juslin, P., Winman, A., & Olsson, H. (2000). Naïve empiricism and dogmatism in confidence research: A critical examination of the hard-easy effect. *Psychological Review, 107*, 384–396.

Kang, S. H. K., McDermott, K. B., & Roediger, H. L., III. (2007). Test format and corrective feedback modify the effect of testing on long-term retention. *European Journal of Cognitive Psychology, 19*, 528–558.

Kearney, E. M., & Zechmeister, E. B. (1989). Judgments of item difficulty by good and poor associative learners. *American Journal of Psychology, 102*, 365–383.

Kelemen, W. L., Frost, P. J., & Weaver, C. A., III. (2000). Individual differences in metacognition: Evidence against a general metacognitive ability. *Memory and Cognition, 28*, 92–107.

Koriat, A. (1997). Monitoring one's own knowledge during study: A cue-utilization approach to judgments of learning. *Journal of Experimental Psychology: General, 126*, 349–370.

Koriat, A., Ma'ayan, H., Sheffer, L., & Bjork, R. A. (2006). Exploring a mnemonic debiasing account of the underconfidence-with-practice effect. *Journal of Experimental Psychology: Learning, Memory, and Cognition, 32*, 595–608.

Koriat, A., Sheffer, L., & Ma'ayan, H. (2002). Comparing objective and subjective learning curves: Judgments of learning exhibit increased underconfidence with practice. *Journal of Experimental Psychology: General, 131*, 147–162.

Lichtenstein, S., & Fischhoff, B. (1977). Do those who know more also know more about how much they know? *Organizational Behavior and Human Performance, 20*, 159–183.

Lichtenstein, S., & Fischhoff, B. (1980). Training for calibration. *Organizational Behavior and Human Performance, 26*, 149–171.

Lovelace, E. A. (1984). Metamemory: Monitoring future recallability during study. *Journal of Experimental Psychology: Learning, Memory, and Cognition, 10*, 756–766.

Maki, R. H., & Berry, S. L. (1984). Metacomprehension of text material. Journal of Experimental Psychology: Learning, Memory. *and Cognition, 10*, 663–679.

Mazzoni, G., & Nelson, T. O. (1995). Judgments of learning are affected by the kind of encoding in ways that cannot be attributed to the level of recall. *Journal of Experimental Psychology: Learning, Memory, and Cognition, 21*, 1263–1274.

McDaniel, M. A., Anderson, J. L., Derbish, M. H., & Morrisette, N. (2007). Testing the testing effect in the classroom. *European Journal of Cognitive Psychology, 19*, 494–513.

Metcalfe, J. (2002). Is study time allocated selectively to a region of proximal learning? *Journal of Experimental Psychology: General, 131*, 349–363.

Metcalfe, J., & Kornell, N. (2003). The dynamics of learning and allocation of study time to a region of proximal learning. *Journal of Experimental Psychology: General, 132*, 530–542.

Metcalfe, J., & Kornell, N. (2005). A region of proximal learning model of study time allocation. *Journal of Memory and Language, 52*, 463–477.

Metcalfe, J., Kornell, N., & Son, L. K. (2007). A cognitive-science based programme to enhance study efficacy in a high- and low-risk setting. *European Journal of Cognitive Psychology, 19*, 743–768.

Nelson, T. O. (1984). A comparison of current measures of the accuracy of feeling-of-knowing predictions. *Psychological Bulletin, 95*, 109–133.

Nelson, T. O., & Dunlosky, J. (1994). Norms of paired-associate recall during multitrial learning of Swahili–English translation equivalents. *Memory, 2*, 325–335.

Nelson, T. O., Dunlosky, J., Graf, A., & Narens, L. (1994). Utilization of metacognitive judgments in the allocation of study during multitrial learning. *Psychological Science, 5*, 207–213.

Pressley, M., & Ghatala, E. S. (1990). Self-regulated learning: Monitoring learning from text. *The Educational Psychologist, 25*, 19–33.

Rawson, K. A., & Dunlosky, J. (2007). Improving students' self-evaluation of learning for key concepts in textbook materials. *European Journal of Cognitive Psychology, 19*, 559–579.

Roediger, H. L., III, & Karpicke, J. D. (2006). Test-enhanced learning: Taking memory tests improves long-term retention. *Psychological Science, 17*, 249–255.

Scheck, P., & Nelson, T. O. (2005). Lack of pervasiveness of the underconfidence-with-practice effect: Boundary conditions and an explanation via anchoring. *Journal of Experimental Psychology: General, 134*, 124–128.

Serra, M. J., & Dunlosky, J. (2005). Does retrieval fluency contribute to the underconfidence-with-practice effect? *Journal of Experimental Psychology: Learning, Memory, and Cognition, 31*, 1258–1266.

Shaughnessy, J. J. (1979). Confidence-judgment accuracy as a predictor of test performance. *Journal of Research in Personality, 13*, 505–514.

Son, L. K., & Metcalfe, J. (2000). Metacognitive and control strategies in study-time allocation. *Journal of Experimental Psychology: Learning, Memory, and Cognition, 26*, 204–221.

Stanovich, K. E. (1986). Matthew effects in reading: Some consequences of individual differences in the acquisition of literacy. *Reading Research Quarterly, 21*, 360–407.

Thiede, K. W. (1999). The importance of accurate monitoring and effective self-regulation during multitrial learning. *Psychonomic Bulletin & Review, 6*, 662–667.

Thompson, W. B., & Mason, S. E. (1996). Instability of individual differences in the association between confidence judgments and memory performance. *Memory and Cognition, 24*, 226–234.

Tobias, S., Everson, H. T., & Laitusis, V. (1999, April). *Towards a performance based measure of metacognitive knowledge monitoring: Relationships with self-reports and behavior ratings*. Paper presented at the annual meeting of American Educational Research Association, Montreal, Canada.

Tuckman, B. W. (1996). The relative effectiveness of incentive motivation and prescribed learning strategy in improving college students' course performance. *Journal of Experimental Education, 64*, 197–210.

Underwood, B. J. (1966). Individual and group predictions of item difficulty for free learning. *Journal of Experimental Psychology, 71*, 673–679.

Vye, N. J., Schwartz, D. L., Bransford, J. D., Barron, B. J., & Zech, L. (1998). SMART environments that support monitoring, reflection, and revision. In D. L. Hacker, J. Dunlosky, & A. C. Graesser (Eds.), *Metacognition in educational theory and practice* (pp. 305–346). Mahwah, NJ: Lawrence Erlbaum Associates, Inc.

Weaver, C. A., III, & Kelemen, W. L. (1997). Judgments of learning at delays: Shifts in response patterns or increased metamemory accuracy? *Psychological Science, 8*, 318–321.

Weaver, C. A., III., Terrell, J. T., Krug, K. S., & Kelemen, W. L. (in press). The delayed JOL effect with very long delays: Evidence from flashbulb memories. In J. Dunlosky & R. A. Bjork (Eds.) *A handbook of memory and metacognition*. Hillsdale, NJ: Lawrence Erlbaum Associates, Inc.

Zechmeister, E. B., Rusch, K. M., & Markell, K. A. (1986). Training college students to assess accurately what they know and don't know. *Human Learning: Journal of Practical Research and Applications, 5*, 3–19.

EUROPEAN JOURNAL OF COGNITIVE PSYCHOLOGY
2007, 19 (4/5), 718–742

How many questions should I answer? Using bias profiles to estimate optimal bias and maximum score on formula-scored tests

Philip A. Higham and Michelle M. Arnold

University of Southampton, Southampton, UK

To maximise multiple choice test scores under formula scoring, students must be able to monitor their knowledge level, reporting correct answers and withholding incorrect ones. Importantly though, neither metacognitive monitoring nor optimal criterion setting has received much attention in the formula-scoring literature. The present research examined the role that both these parameters have on obtained test score. Students wrote three exams under formula-scoring rules, and from these data, we created bias profiles using methodology developed by Higham (2007), which allowed us to estimate the optimal number of questions that students needed to answer to maximise their scores. The results indicated that higher scoring students monitored their knowledge best, but that all students tended to guess too often, lowering their test scores. This conservatism maintained despite feedback from previous tests. The results also showed that monitoring is a stable participant skill that generalises across tests.

Multiple choice examinations are popular in classroom settings because they measure knowledge and understanding of a particular topic with minimal effort, even if administered to large classes. However, because a significant number of correct answers on such tests can be selected without any knowledge or understanding, (simply by choosing a response at random), the issue of how best to score the test so that it is reflective of true knowledge is not so straightforward. In some cases, a "correction for guessing" or *formula scoring* is applied, whereby correct responses score points, but errors are penalised (e.g., Holzinger, 1924; Muijtjens, van Mameren, Hoogenboom, Evers, & van der Vleuten, 1999; Thurstone, 1919). To avoid the penalty, students can opt to "pass", which typically results in neither a gain nor loss of points (although see Traub, Hambleton, & Singh, 1969, for an exception).

Correspondence should be addressed to Dr Philip A. Higham, School of Psychology, University of Southampton, Highfield, Southampton SO17 1BJ, UK. E-mail: higham@soton.ac.uk

Decisions about which and how many questions to omit involves the strategic regulation of accuracy (e.g., Koriat & Goldsmith, 1996). That is, students must monitor their own level of knowledge so that they omit answers that likely are wrong, and respond with answers that likely are correct. Students with excellent monitoring may not necessarily lose many points with formula scoring because they will omit all their incorrect responses (however many of those there may be), thus avoiding the penalty. Conversely, students with poor monitoring may offer several incorrect responses, erroneously believing them to be correct, which would cause a detriment to their overall score. Nevertheless, despite the clear influence that metacognitive monitoring ability has on the score obtained with formula scoring, very little research has focused on (or even attempted to measure) metacognition in the context of formula-scored tests (although see Higham, 2007). In this paper we have attempted to fill that gap by using a framework that is based on signal detection theory (SDT).

A SIGNAL DETECTION FRAMEWORK OF THE STRATEGIC REGULATION OF ACCURACY

Recently, Higham (in press) applied an SDT framework to performance on the Scholastic Aptitude Test (SAT), a formula-scored test written by millions of American high-school seniors each year as an entrance requirement for university. According to this framework, students' own correct and incorrect responses to test questions define the signal-plus-noise and noise trials, respectively, and the discrimination task is to distinguish between these two types of response.[1] To make this discrimination, it can be assumed that a monitoring mechanism assigns confidence to candidate responses (i.e., question alternatives) and a malleable criterion is set somewhere on an underlying subjective confidence-in-accuracy dimension (see also Koriat & Goldsmith, 1996).[2] If the highest level of confidence associated with one of the alternatives for a given test question is at or above the criterion level of confidence, it is judged "correct" and reported. Otherwise, it is judged "incorrect" and withheld.

---

[1] Technically, such discrimination is known as response-contingent, or Type 2 SDT. For further description and discussion regarding Type 2 SDT, see Banks (1970), Clarke, Birdsall, and Tanner (1959), Galvin, Podd, Drga, and Whitmore (2003), and Higham (2002, 2007).

[2] We make no assumptions about the monitoring mechanism at this stage other than it is used to assign confidence to the question alternatives. For example, the mechanism might operate consciously or unconsciously; similarly, the assignment process might be exhaustive or terminate once a candidate of reportable confidence has been found. These are important details that we will need to ascertain at some point in the future, but they are not critical to the current research.

For most tests involving formula scoring, the withheld answers are not known. However, a few studies have required participants to go back after finishing the test to provide best guesses to questions initially omitted (e.g., Bliss, 1980; Cross & Frary, 1977; Ebel, 1968; Higham, 2007, Exp. 1; Muijtjens et al., 1999; Sax & Collet, 1968; Sherriffs & Boomer, 1954; Slakter, 1968a, 1968b). The focus of many of these studies has been on ascertaining whether the accuracy of withheld answers is above chance and, indeed, in most cases, it is. However, in terms of the SDT framework, the reason for determining the number of correct and incorrect withheld answers is so that the hit (H) and false alarm (FA) rates can be calculated. In particular, after completing the whole test, the proportions of all the correct responses and all the incorrect responses that are judged "correct" define the H and FA rates, respectively. From the H and FA rates, it is possible to then calculate a discrimination index (e.g., d'), which estimates metacognitive monitoring. High levels of discrimination indicate that the monitoring mechanism is assigning confidence appropriately, leading to a high confidence–accuracy correlation. A bias index (e.g., β) can also be derived from the H and FA rates: This index estimates the tendency towards making "correct" judgements, or in the context of formula-scored tests, the tendency to report answers. The basic SDT model is shown in Figure 1, with the assumption that the distributions of correct and incorrect items are normally distributed over confidence and have equal variance.

The results from the two-stage testing paradigm in which participants are required to provide best guesses to items originally omitted have been

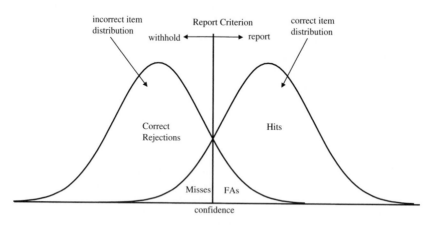

Figure 1. Equal variance, Gaussian, Type 2, signal detection model. The criterion in the figure is unbiased. Liberal and conservative bias would be indicated if the criterion appeared to the left or to the right of the point at which the two curves intersect, respectively.

criticised (e.g., see Lord, 1975). For example, during the second fill-in-the-blanks stage, "don't know" questions are given additional processing time and it is possible that, with this extra time, participants are able to recall the correct responses from memory. Hence, as an alternative methodology, Higham (2007, Exp. 2) had participants answer each question only once, but they were required to place their responses in one of two categories. In the "go for points" category, correct answers resulted in points and incorrect answers garnered a penalty; these responses were considered analogous to reported (high confidence) answers. In the "guess" category, points were neither gained nor lost for correct or incorrect answers, and these responses were considered analogous to withheld (low confidence) answers. Because this procedure avoids the pitfalls of the two-stage testing paradigm, it was adopted in the current study.

As noted above, monitoring has a clear effect on the corrected score, but so does report bias (the other SDT parameter). Report bias is an important consideration for students because it means that their test-taking strategy partially determines their final score. Assuming reasonable monitoring levels, if too few questions are answered (conservative criterion), there will be an opportunity cost because many correct candidates will be withheld. On the other hand, answering too many questions (liberal criterion) will result in many incorrect answers incurring the penalty (see Figure 1). Nonetheless, just as with monitoring, sophisticated treatment of the influence of report bias on formula-scored tests been lacking in the testing literature, possibly because the tools necessary to determine its effect have not been available. The SDT framework provides those tools and allows report bias and metacognitive monitoring to be estimated separately so that their individual contributions to the corrected score can be determined.

Higham (2007) argued that the extent to which conservative versus liberal responding affects the corrected score is dependent on the combination of students' ability to monitor their knowledge level, the penalty assigned to errors, and the raw score (i.e., the sheer number of correct responses after all questions have been answered, uncorrected by penalties). From the students' perspective, it is desirable to balance all of these factors and set a criterion that will maximise their corrected score—a criterion setting that we refer to as *optimal bias*. However, optimal bias is not easily determinable even for sophisticated students because the level of one variable that is associated with the maximum score is dependent on the levels of the others. Nonetheless, it is possible to use the three factors identified by Higham to generate a *bias profile* that can be used to predict each individual's optimal level of bias.

DETERMINING OPTIMAL BIAS FROM BIAS PROFILES

A bias profile is a plot of the corrected score as a function of the proportion of (low-confidence) responses assigned to the "guess" category, an example of which is shown in Figure 2. For generalisations beyond the particular two-category response assignment methodology used here and in Higham (2007), the x-axis could be labelled "proportion omissions" or "proportion withheld", as these are considered conceptually equivalent to "proportion guesses" in this context. However, because we have adopted Higham's two-category methodology, there are no "omitted" or "withheld" responses in the strict sense of those terms (i.e., participants must always choose a response). Instead, the categories "answer" and "guess" were used to denote the high and low confidence categories, respectively, and so these terms will be used in place of "report" and "withhold" for the remainder of this paper.

At a minimum, the parameters needed to generate a bias profile include the metacognitive monitoring level (d'), the overall uncorrected raw score on the test after pooling responses from both the "answer" and "guess" categories (f), and the penalty for errors (p). The mathematical details of how bias profiles can be generated from these parameters are shown in the first part of the Appendix. In the research described above in which the low confidence answers were obtained using either a two-stage testing paradigm or Higham's (2007) two-category methodology, neither correct nor incorrect

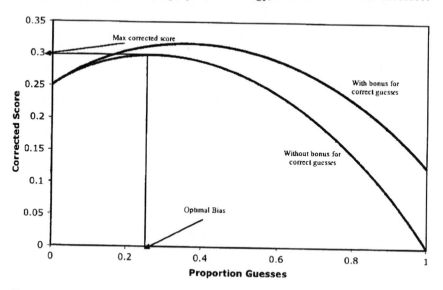

Figure 2. Bias profile for $d' = 1.00$, f $= 0.50$, p $= 0.50$ and b $= 0$ (bottom line) and $d' = 1.00$, f $= 0.50$, p $= 0.50$ and b $= 0.25$ (top line). Maximum corrected score $= 0.30$. Proportion of blanks that yields maximum corrected score (optimal bias) $= 0.27$. All parameters are defined in the text.

guesses were associated with any points. However, with the current tests, a small bonus (b = +0.25) was offered for correct responses in the "guess" column. If included as part of the test, this bonus parameter must be incorporated into the calculation of the bias profile, the mathematical details of which are shown towards the end of Appendix. For tests in which no bonus is offered, the parameter b is simply set to zero and can be ignored.

It is possible to get an intuitive sense of bias profiles by examining Figure 2. For the moment, consider only the bottom curve of the two that are plotted, the one marked "without bonus for correct guesses". This particular profile was generated for someone with moderate monitoring ($d' = 1.00$), a moderate overall score (f = 0.50), and a moderate-to-high penalty for errors (p = 0.50). At the extreme right of the profile (the point at which all answers were guesses), the corrected score is zero because no points were gained or lost. At the extreme left of the profile, no answers are guesses, so the corrected score (0.25) is equal to the raw proportion correct (f = 0.50) minus one half (p = 0.50) of the raw proportion incorrect ([$1-f$] = 0.50). In between these two extremes, the curve rises to a maximum of 0.30, which corresponds to 0.27 of items being guessed, and then drops off again. Thus, the bias profile for this individual indicates that his/her optimal bias is equal to 0.27 (i.e., 27% of the items on the test should be placed in the "guess" category), which will achieve a maximum corrected score of 0.30. This point is indicated in Figure 2.

If a bonus for correct guesses is used, as it is in the current study, it will lift the right hand portion of the curve, where guessing is more common, changing both the optimal bias and the maximum corrected score that is obtainable. The upper curve in Figure 2 is the bias profile that is matched on all parameters used to generate the other profile plotted in the figure, only with a bonus of 0.25 points for each correct guess. Most intuitively, in the case of extreme conservatism (guessing all responses) the corrected score is no longer equal to zero, but instead is dependent on the size of the bonus and the number of correct guesses. For the example in Figure 2, the corrected score associated with extreme conservatism is equal to 0.125, which is obtained by earning 0.25 points for each guess that was correct (0.50).[3]

It is important to note that these profiles rely on some assumptions. In particular, they assume that the correct and incorrect items are distributed normally over confidence and that they have equal variance, similar to the distributions shown in Figure 1. Using data from Receiver Operating Characteristic (ROC) curves, Higham (2007) demonstrated that a model

---

[3] Although a bonus was offered for correct guesses in the present study, the pattern of results is very similar to Higham's (2007), in which no such bonus was offered. Hence, we do not believe that the results obtained in the current research are idiosyncratic.

such as this was able to predict performance on formula-scored tests very well. Consequently, similar assumptions are adopted for the current study.

OVERVIEW OF THE STUDY AND RESEARCH QUESTIONS

The data for the current study were obtained from three examinations in a classroom setting for which the outcome had real-world consequences (i.e., as the basis for the final grade/standing in the course). Data from two midterms and a final examination were analysed, all of which were four-alternative, multiple choice tests with a penalty for errors. Following Higham (2007), participants indicated their confidence in the accuracy of their responses by choosing to respond on the test sheet in either the "answer" column (high confidence response which could incur a penalty) or the "guess" column (low confidence response with no penalty possible).

These examinations were administered to a first-year class of university students who were taking Introductory Psychology as an optional course. Because of the unique scoring method used with the examinations (detailed below), it was possible to calculate monitoring (d') and the overall uncorrected score (f), hereafter referred to as the "raw score". This information coupled with the test parameters allowed bias profiles to be generated, which were, in turn, used to determine optimal bias. A comparison was then made between the corrected score that students actually obtained with the corrected score that could have been obtained if bias had been set optimally. In other words, we addressed the question, do strategic factors, such as the tendency to risk the penalty on too many/few items, have any effect on the score that is obtained and, if so, how far is the obtained score from the optimal score? An additional interest was in the effect of feedback; that is, do students learn from their mistakes? If the obtained score suffers as a result of nonoptimal bias setting on Test 1, is the bias adjusted to be more optimal on subsequent tests? Finally, because each student wrote three tests, we were able to determine whether both monitoring and the raw score were reliable across tests; if so, it is conceivable that bias profiles are stable within individuals, and therefore they can be used to predict optimal bias across different tests.

METHOD

Participants

The participants were 59 students who completed all three exams in an introductory psychology course at the University of Southampton. All students were majoring in another discipline and were taking Psychology as a first year optional unit.

Design and materials

All three exams consisted of questions that were based on the course textbook and lecture notes; there were 60 questions on both Test 1 and Test 2, and 100 questions on Test 3. All questions were paired with four possible alternatives, which consisted of the correct answer and three plausible incorrect responses. For example, a typical question might be "Iconic memory corresponds to which of our senses? (a) vision (b) audition (c) olfaction (d) gustation", for which the correct answer is "a". The answer sheet for the questions contained two separate columns of four responses (a, b, c, and d), with one column designated "answer" and the other column designated "guess". For each category of response, the points gained or lost for correct and incorrect answers was indicated. For example, for Test 1, "($+1$, -0.5)" appeared next to "Answer", whereas "($+0.25$, 0)" appeared next to "Guess".

In terms of SDT, responding with a correct answer in the "answer" section was defined as a hit (H), whereas responding with a correct answer in the "guess" section was designated a miss (M). In a similar vein, an incorrect answer in the "answer" column was defined as a false alarm (FA), but in the "guess" section, it was defined as a correct rejection (CR). Across all three tests, Hs, Ms, and CRs were always worth 1.00, 0.25, and 0 points, respectively. In Test 1, a FA resulted in a penalty of 0.50, and in Test 2 and 3, a penalty of 0.25. Some readers may have noticed that the penalties used do not correspond to the standard penalty typically associated with formula scoring (i.e., a penalty of 0.33 for a test with four alternatives). The standard penalty is used to correct for blind guessing (i.e., choosing an alternative at random if the answer is not known). For the current tests, the penalty was used to lower the test mark average so that it was comparable to the average mark in other classes where other assessment formats were used (e.g., essay examinations), not to correct for random guessing. Thus, there was no reason to adhere to the standard penalty.

The total scores were divided by the number of items on the tests to yield rates. Thus, the corrected score for Test 1 was $(1.00)HR + (0.25)MR - (0.50)FAR$, whereas the raw score was simply $HR + MR$, where HR, FAR, and MR designate the H rate, FA rate, and M rate, respectively. Metacognitive monitoring was measured with d', which is the distance between the means of the correct and incorrect item distributions in standard deviation units. Bias was measured by the proportion of guesses on the test (i.e., responses in the "guess" column); for example, more guessing amounted to a more conservative response criterion. One interpretation of criterion setting is that students were "unbiased" if they guessed 50% of the time. However, we are most interested in the degree to which participants deviate from optimal bias and the effect this has on the corrected test score. Hence, we reserve the terms "conservative" and "liberal" to respectively refer to cases in which guessing occurred too often, or too infrequently, to achieve the maximum corrected score.

Procedure

Participants completed all three exams during scheduled class meetings. Although the penalty for an incorrect response differed between Test 1 on the one hand and Tests 2 and 3 on the other, the points instructions were consistent across all three tests. Participants were told that they had two options when answering a question, in that they could choose to "answer" or "guess". Specifically, they were informed that if they believed that they knew the correct answer to a question then they should place their response in the "answer" section; if they circled the correct answer they would receive +1 point, but if they gave an incorrect response they would be penalised −0.50 (Test 1) or −0.25 (Test 2 and 3). Conversely, if they believed they did not know the correct answer to a question (i.e., that they were guessing) then they were told to respond in the "guess" column, in which a correct answer earned them +0.25 and an incorrect response resulted in 0 points.

The following week, the correct answers to all questions on Tests 1 and 2 were provided. This feedback occurred during a regular class meeting in which the instructor handed back answer sheets to students, displayed the questions and answer alternatives to the class, and then indicated which alternative was correct on a question-by-question basis. No feedback pertaining specifically to responding strategy for answering and guessing was given to the students, although they did have all the necessary information to work out for themselves their different scoring scenarios (e.g., the score they would have obtained if they had answered the questions for which they had provided correct guesses).

RESULTS AND DISCUSSION

For the following analyses, "corrected score" refers to the score on the test after penalties and bonuses have been subtracted and added, respectively, to the number right in the "answer" column, whereas "raw score" refers to the total number right in both the "answer" and "guess" columns. Both the raw and corrected scores were divided by the number of questions on the test so that the values ranged between zero and one and were comparable between tests. Both scores are reported in Table 1 for each of the three exams. A 3 (test: Test 1, Test 2, Test 3) × 2 (score: raw, corrected) within-subjects analysis of variance (ANOVA) on the obtained scores revealed a main effect of score type, $F(1, 58) = 650.71$, $MSE = 0.003$, $\eta_p^2 = .92$, $p < .001$, and a main effect of test, $F(2, 116) = 37.06$, $MSE = 0.02$, $\eta_p^2 = .39$, $p < .001$. As shown in Table 1, Test 1 ($M = 0.53$, $SEM = 0.02$) was more difficult than both Test 2 ($M = 0.67$, $SEM = 0.02$), $t(58) = 7.87$, $p < .001$, and Test 3 ($M = 0.62$, $SEM = 0.02$), $t(58) = 6.43$, $p < .001$, and Test 3 was more difficult than Test 2, $t(58) = 2.65$, $p = .01$. Additionally, as expected, the overall corrected score ($M = 0.54$, $SEM = 0.02$) was reliably lower than the raw score ($M = 0.67$, $SEM = 0.02$). There was also an interaction between score type and test, $F(2, 116) = 71.44$, $MSE = 0.001$, $\eta_p^2 = .55$, $p < .001$, which occurred because the difference between the raw and corrected scores was significantly greater for Test 1 ($M = 0.19$, $SEM = 0.01$) than either Test 2 ($M = 0.11$, $SEM = 0.01$), $t(58) = 9.68$, $p < .001$, or Test 3 ($M = 0.12$, $SEM = 0.01$), $t(58) = 8.87$, $p < .001$; however, there was no reliable difference in the size of the score type effect between Test 2 and Test 3, $t(58) = 1.92$, $p = .06$. This larger difference between the scores for the first test is likely due to the fact that it had the higher penalty for errors (0.50), and was a more difficult test than either of the other tests (both 0.25).

Before conducting further analyses, a rank variable was constructed by averaging the raw scores from all three tests and splitting the participants into the top third ($n = 20$, $M = 0.80$, $SEM = 0.01$), the middle third ($n = 19$, $M = 0.69$, $SEM = 0.01$), and the bottom third ($n = 20$, $M = 0.53$, $SEM = 0.02$). The average scores for monitoring (d'), split both by rank and test, are reported in Table 2. The monitoring data were analysed in a 3 (test: Test 1, Test 2, Test 3) × 3 (rank: top, middle, bottom) mixed ANOVA, with rank as the between-subjects variable. The analysis revealed a main effect of rank, $F(2, 56) = 15.49$, $MSE = 0.35$, $\eta_p^2 = .36$, $p < .001$, but no main effect of test, nor a significant interaction, both $Fs < 1$. The main effect of rank occurred because students in the top third of the class had better monitoring ($M = 1.26$, $SEM = 0.10$) than students in the middle third ($M = 0.94$, $SEM = 0.06$), who, in turn, had better monitoring than students in the bottom rank ($M = 0.66$, $SEM = 0.07$). Post hoc Tukey HSD tests revealed that all means were significantly different from all others, all $ps \leq .04$. These results suggest that monitoring ability covaries with the score obtained on

TABLE 1
Mean raw and corrected scores across the three exams

Score	Test 1	Test 2	Test 3
Raw[a]	0.62 (0.020)	0.72 (0.016)	0.68 (0.019)
Corrected[b]	0.43 (0.026)	0.61 (0.020)	0.56 (0.023)

Standard errors of the means are in parentheses.
[a]Raw $= HR + MR$, where HR and MR represent the hit and miss rates, respectively.
[b]Corrected score on Test 1 $= HR + (-.50)FAR + (.25)MR$; Corrected score on Tests 2 and 3 $= HR + (-.25)FAR + (.25)MR$, where FAR represents the false alarm rate.

TABLE 2
Mean monitoring (d') on the three tests as a function of rank

Rank	Test 1	Test 2	Test 3
Top third	1.20 (0.11)	1.29 (0.12)	1.29 (0.13)
Middle third	0.93 (0.10)	0.99 (0.13)	0.88 (0.08)
Bottom third	0.61 (0.12)	0.64 (0.11)	0.73 (0.07)

Standard errors of the means are in parentheses.

the test; compared to poorer students, the students who know the material being tested are also better at monitoring what they know. This result is analogous to that obtained in other research in metacognition (e.g., Perfect & Stollery, 1993).

To investigate criterion setting, bias profiles were generated for each participant and, from these, individual optimal bias scores (the proportion of guesses that maximises the corrected test score) were determined. Mean optimal bias scores, mean actual bias scores (i.e., the actual proportion of "guess" responses), and the mean absolute deviation of actual bias from optimal bias, are displayed in Table 3. The guess proportion data were then analysed in a 2 (bias type: optimal, actual) × 3 (test: Test 1, Test 2, Test 3) × 3 (rank: top, middle, bottom) mixed-design ANOVA, with test and bias type as within-subject factors and rank as the between-subjects factor. The analysis revealed a main effect of test, $F(2, 112) = 74.80$, $MSE = 0.01$, $\eta_p^2 = .57$, $p < .001$, bias type, $F(1, 56) = 95.67$, $MSE = 0.02$, $\eta_p^2 = .63$, $p < .001$, and rank, $F(2, 56) = 26.44$, $MSE = 0.03$, $\eta_p^2 = .49$, $p < .001$. The main effect of test arose because both actual and optimal responding on Test 1 ($M = 0.28$, $SEM = 0.02$), which had a higher penalty for errors, was more conservative than on either Test 2 ($M = 0.12$, $SEM = 0.01$), $t(58) = 7.73$, $p < .001$, or Test 3 ($M = 0.15$, $SEM = 0.01$), $t(58) = 8.33$, $p < .001$. The bias type main effect occurred because, overall, actual bias ($M = 0.25$, $SEM = 0.02$) was too conservative relative to optimal bias ($M = 0.10$, $SEM = 0.01$). The main effect of rank was due to students in the bottom third ($M = 0.26$, $SEM = 0.02$) having higher bias scores than students in the top ($M = 0.12$, $SEM = 0.01$) and middle ($M = 0.16$, $SEM = 0.01$) thirds of the class, Tukey's HSD, $ps < .001$. The analysis also revealed an interaction between test and rank, $F(4, 112) = 11.49$, $MSE = 0.01$, $\eta_p^2 = .29$, $p < .001$, which was qualified by a significant three-way interaction, $F(4, 112) = 7.12$, $MSE = 0.01$, $\eta_p^2 = .20$, $p < .001$. Follow-up analyses demonstrated that the three-way interaction was due to a different pattern of results for actual and optimal bias in the bottom rank of the class. Specifically, actual bias was always significantly greater than optimal bias across all three exams for students in the top and middle thirds of the class, all $ts \geq 6.05$, $ps < .001$. For students in the bottom third of the class, actual bias also was reliably greater than optimal bias in Tests 2 and 3, $ts \geq 7.23$, $ps < .001$; however, there was no difference between their actual and optimal bias in Test 1, $t(19) = 0.23$, $p = .82$. No other effects from the ANOVA were significant.

These results indicate that, generally speaking, students were too conservative in their responding, in that they were guessing too often (see Table 3 and Figures 3–5). The one exception to this general rule was the lower third of the class who seemingly chose a criterion very close to optimal on the first test, at least as suggested by a comparison of mean optimal and actual bias. However, this result needs to be treated with caution. As shown in Figure 5, the distribution of deviations from optimal bias for the lowest third of the class was widely dispersed, with a mean close to zero. Thus, although the lower third group as a whole appeared to be setting an optimal criterion on the first test, closer analysis of the data reveals that, in fact, any given student's criterion setting was likely to be way off.

A different picture of the gap between actual and optimal bias emerges if, instead of examining the *signed* deviation, the *absolute* deviation from optimal bias is analysed. It is important also to look at the absolute deviation because simply looking at the signed deviation within-groups can be deceptive; if some participants are too conservative and others too liberal then the deviation scores may cancel each other out and it would appear as though the group were performing close to optimal bias when if fact they were not. A 3 (test: Test 1, Test 2, Test 3) × 3 (rank: top, middle, bottom) ANOVA conducted on the absolute deviation from optimal bias showed a main effect of rank, $F(2, 56) = 8.55$, $MSE = 0.03$, $\eta_p^2 = .23$, $p = .001$. Contrary to the previous analysis on signed deviation from optimal bias, post hoc Tukey HSD tests revealed that the absolute deviation from optimal bias for the bottom third of the class ($M = 0.25$, $SEM = 0.03$) was *greater* than either the middle third ($M = 0.16$, $SEM = 0.02$), $p < .01$, or the top third ($M = 0.14$, $SEM = 0.02$), $p = .001$, whereas the middle and top thirds did not differ, $p = .83$. Neither the main effect of test nor the interaction was significant, $F(2, 112) = 2.29$, $p = .11$, and $F < 1$, respectively. These results, along with the data shown in Figure 5, suggest that the poorest students on the first test seemed to be setting criteria almost randomly, leading to high deviations from optimal bias, but without a particular leaning toward liberalism or conservatism. On the second and third test, the poorest students tended towards conservatism, just as other students did, but their deviation from optimal bias was greater. The absence of an effect of test on the absolute deviation from optimal bias suggests that students did not learn from feedback on previous tests regarding how to optimise their test-taking strategy.

What impact did nonoptimal criterion setting actually have on the corrected scores that students obtained? To answer this question, the actual

TABLE 3

Mean optimal bias (proportion of answers students should have guessed) and actual bias (proportion of answers students did guess) by rank (top, middle, and bottom third of class) across the three exams

Bias type	Top			Middle			Bottom		
	T1	T2	T3	T1	T2	T3	T1	T2	T3
Optimal	0.08 (0.01)	0.02 (0.004)	0.03 (0.01)	0.15 (0.03)	0.05 (0.01)	0.03 (0.01)	0.44 (0.07)	0.04 (0.02)	0.08 (0.02)
Actual	0.24 (0.02)	0.15 (0.02)	0.17 (0.02)	0.31 (0.03)	0.18 (0.02)	0.21 (02)	0.42 (0.05)	0.25 (0.03)	0.34 (0.04)
DOB[a]	0.15 (0.02)	0.13 (0.02)	0.15 (0.02)	0.16 (0.03)	0.14 (0.02)	0.18 (0.02)	0.29 (0.06)	0.21 (0.03)	0.26 (0.03)

[a]DOB = the mean of the *absolute value* of the deviation of actual bias from optimal bias. Standard errors of the means are in parentheses.

TABLE 4

Mean maximum possible score (corrected score students would have received if performing at optimal bias) and actual corrected score (corrected score students did receive) by rank (top, middle, and bottom third of class) across the three exams

Score type	Top			Middle			Bottom		
	T1	T2	T3	T1	T2	T3	T1	T2	T3
Maximum	0.66 (0.02)	0.79 (0.02)	0.76 (0.02)	0.48 (0.03)	0.66 (0.03)	0.62 (0.02)	0.25 (0.03)	0.51 (0.03)	0.42 (0.03)
Actual	0.62 (0.02)	0.75 (0.02)	0.72 (0.02)	0.45 (0.03)	0.62 (0.02)	0.58 (0.02)	0.22 (0.03)	0.47 (0.03)	0.39 (0.03)
DOS[a]	0.04 (0.01)	0.04 (0.01)	0.04 (0.01)	0.03 (0.01)	0.04 (0.01)	0.04 (0.01)	0.03 (0.01)	0.04 (0.01)	0.04 (0.01)

[a]DOS = the mean deviation of actual score from optimal score. The difference between the maximum possible and actual corrected scores across the three tests for each rank are all significant at the .01 level.

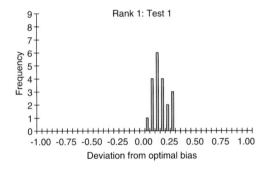

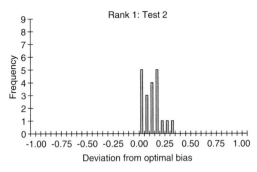

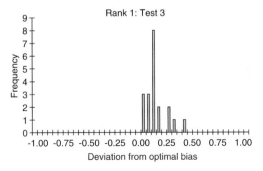

Figure 3.   The frequency distribution for deviation from optimal bias scores across Tests 1, 2, and 3 for students in the top third of the class (rank 1). Scores to the right and left of zero indicate conservative and liberal bias, respectively.

corrected score for each student was compared to the score that was theoretically possible if the report criterion had been set optimally. The latter score was gleaned from the bias profile of each student, and both the mean maximum corrected score and the mean actual corrected score are reported in Table 4. The corrected score data were analysed in a 2 (score type: maximum, actual) × 3 (test: Test 1, Test 2, Test 3) × 3 (rank: top, middle, bottom) mixed-design ANOVA, with test and score type as within-subject factors and rank as

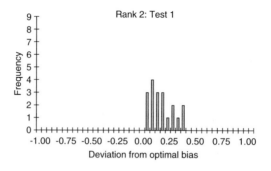

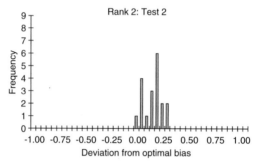

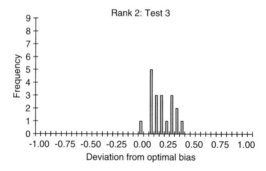

Figure 4.   The frequency distribution for deviation from optimal bias scores across Tests 1, 2, and 3 for students in the middle third of the class (rank 2). Scores to the right and left of zero indicate conservative and liberal bias, respectively.

the between-subjects factor. There was a main effect of test, $F(2, 112) = 53.55$, $MSE = 0.02$, $\eta_p^2 = .49$, $p < .001$ (Test 1: $M = 0.45$, $SEM = 0.03$; Test 2: $M = 0.63$, $SEM = 0.02$; Test 3: $M = 0.58$, $SEM = 0.02$), score type, $F(1, 56) = 171.63$, $MSE = 0.001$, $\eta_p^2 = .75$, $p < .001$ (maximum: $M = 0.57$, $SEM = 0.02$; actual: $M = 0.54$, $SEM = 0.02$), and rank, $F(2, 112) = 115.32$, $MSE = 0.03$, $\eta_p^2 = .81$, $p < .001$ (rank 1: $M = 0.72$, $SEM = 0.01$; rank 2: $M = 0.57$, $SEM = 0.01$; rank 3: $M = 0.38$, $SEM = 0.02$). There was no

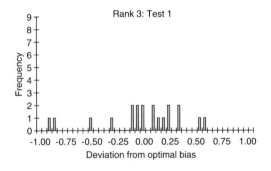

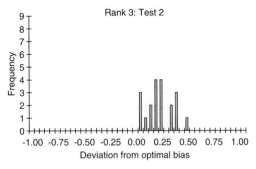

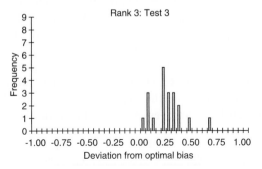

Figure 5. The frequency distribution for deviation from optimal bias scores across Tests 1, 2, and 3 for students in the bottom third of the class (rank 3). Scores to the right and left of zero indicate conservative and liberal bias, respectively.

interaction between test and rank, $F(4, 112) = 2.29$, $p = .06$, test and score type, $F(2, 112) = 1.96$, $p = .15$, or score type and rank, $F < 1$, or between test, score type, and rank, $F < 1$. Thus, all students seemed to score lower than they could have given their knowledge and monitoring ability. The absence of a rank by score type interaction is important because it indicates that students of all ability levels were affected about equally by failing to adopt a criterion that was optimal.

Finally, because the same students sat three tests, we were able to investigate the consistency of some of the performance measures across tests. Indeed, if bias profiles are to be used as predictors of future performance, then it is important to demonstrate that those aspects of participants' performance that affect bias profiles are reliable at least somewhat across tests. Specifically, two measures of performance are critical; raw score and monitoring, as both of these are parameters used directly to generate bias profiles (see Appendix for details). Although not a parameter, the corrected score was included in the following analysis for completeness. The Pearson correlation coefficients between the raw scores across tests, and between the raw score and the corrected scores, are presented in Table 5. All correlations were significantly above zero both within each test (e.g., raw score versus corrected score for Test 1) and across the tests (e.g., raw score for Test 1 and raw score for Test 2), all $ps \leq .01$. Similarly, for monitoring (d'), the correlations between Test 1 and Test 2 ($r = .29$), Test 1 and Test 3 ($r = .41$), and Test 2 and Test 3 ($r = .48$), were all significantly above zero, all $ps \leq .03$. Furthermore, mean monitoring remained remarkably stable across tests (Table 2), despite differences in the level of difficulty of these tests and variations in the penalty for errors. Together, these results demonstrate that both the raw and corrected scores on the one hand, and monitoring ability on the other, are relatively stable participant variables across tests, and suggest that a bias profile generated by one test can be used to predict optimal performance on another related test.

The consistency of the bias profiles across tests for groups of students as whole can be seen in Figure 6, which shows the bias profiles for the top, middle, and bottom thirds of the class for each of the three tests. The consistency between Test 2 and Test 3, which shared the 0.25 penalty, is most

TABLE 5
Correlations between the raw and corrected score across the three tests

| | Raw score | | |
Score type	Test 1	Test 2	Test 3
Raw score			
Test 1		.61	.79
Test 2			.57
Corrected score			
Test 1	.97	.62	.77
Test 2	.65	.97	.64
Test 3	.76	.56	.98

All of the correlations are significant at the .01 level.

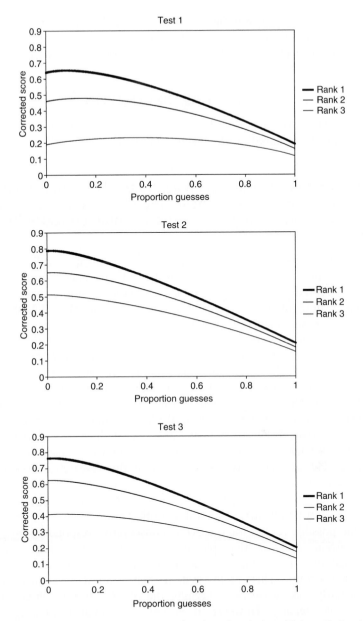

Figure 6. Bias profiles for the three tests as a function of rank (top third = rank 1; middle third = rank 2; bottom third = rank 3).

striking. It is also worth noting how the corrected-score differences between the three rank levels are more pronounced if few guesses are provided, and how optimal bias varies with rank (see also Table 3).

GENERAL DISCUSSION

In this section we address each of our research questions in turn, providing answers to them based on the results obtained.

1. Do strategic factors, such as the tendency risk the penalty on too many/few items, have any effect on the score that is obtained and, if so, how far is the obtained score from the optimal score?

The answer to the first part of this question is unequivocally "yes". The results from the present study demonstrate that students writing genuine formula-scored tests in a university classroom setting have difficulty deciding how often to risk the penalty to gain the highest possible corrected score. Generally speaking, students avoided the penalty and placed responses in the "guess" category too often, and this behaviour resulted in an opportunity cost because the accuracy of these "guess" responses was high. Specifically, the accuracy of "guess" responses for the top third of the class was 0.47, 0.47, and 0.47 for Tests 1, 2, and 3, respectively. The analogous values for the middle third were 0.41, 0.42, and 0.43, and for the bottom third, 0.33, 0.43, and 0.34, respectively. Had these responses been offered as "answers" (which risked the penalty) instead of "guesses", the corrected test score would have increased on average. The answer to the second part of this research question is that there was a small but consistent loss of points on the tests due to nonoptimal bias setting (about 4%; see Table 4). This opportunity cost was consistent both across tests and rank groups.

One possible reason for the overly conservative responding observed in the current study is that placing responses in the "guess" category provided the opportunity of a small bonus for correct answers, but no loss for errors. This point system may appeal to many students, particularly risk-averse ones, because the other "answer" choice involved a possible penalty. However, it is not a rational test-taking strategy because there is an overestimation of the possible points to be gained by "guessing" and/or an overestimation of the impact of the penalty for errors by "answering". Support for the latter type of overestimation comes from students who wrote the SAT in Higham's (2007) research; no bonus for correct "guessing" was offered, yet these students also tended to place too many answers in the "guess" category, which hurt their corrected score.

Results from Higham's (2007) first experiment also demonstrate that overly conservative responding is not a function of the two-category "guess/answer" response methodology that we have advocated. In that experiment, participants returned to questions for which answers were omitted on the first pass through the test and provided "best guesses" without a penalty system in place. The accuracy of this pool of "best guesses" was also sufficiently high to more than compensate for any penalties that would have been incurred had the questions been answered on the first pass. Overall, the results from the current research, coupled with the data in Higham, suggest that conservative responding on formula-scored tests is a fairly general phenomenon and not specific to particular materials, to cases in which bonuses are given for correct guesses, or to specific methodologies used to separate high and low confident responses.

2. What is the effect of feedback; do students learn from their mistakes? If the obtained score suffers as a result of nonoptimal bias setting on Test 1, is the bias adjusted to be more optimal on subsequent tests?

The data from the present study indicated that students do not appear to learn from their mistakes or from the feedback given to them. Repeated testing with formula scoring did not teach them the optimal balance between "answer" and "guess" responses for their knowledge level and the characteristics of the current test. Consequently, students not only failed to reach optimal bias by the final exam, their deviation from optimal bias remained almost invariant across the different tests (see Table 4). These results are consistent with Higham (2007), who found that even giving students trial-by-trial feedback (i.e., showing the overall score and the points won/lost on the previous trial) had no effect on performance.

3. How reliable is the raw score and monitoring across different tests?

The answer to this question appears to be "somewhat". Correlations between Test 1 and Test 2, Test 1 and Test 3, and Test 2 and Test 3 were significantly above zero for both monitoring and the raw score. Thus, students who knew a lot, or who monitored their knowledge well, on one test were likely to know a lot and monitor their knowledge well on another. However, it is worth noting that some of the correlations were not high: For example, the monitoring correlation between Test 1 and Test 2 was only $r = .29$. The reliability of monitoring is an important issue because the extent to which a bias profile generated from performance on one test can be used to guide optimal bias setting on a subsequent test is dependent on the underlying parameters remaining constant. These parameters include monitoring, the raw score, and the penalty for errors. Of course, the penalty and level of difficulty of the test changed between Test 1 and Test 2, which could partially account for the particularly low monitoring correlation between those tests. Nonetheless, the fact that both monitoring and the raw score are not test-specific parameters provides some hope that bias profiles and the data they generate can be used to make between-test predictions. Thus, students about to write the SAT could practise on previous versions of the SAT, generate their bias profile, and predict optimal bias for the actual SAT. (See also Keleman, Winningham, & Weaver, 2007 this issue, regarding the benefits of repeated testing on SAT scores, and Rawson & Dunlosky, 2007 this issue, concerning improvements due to a "restudy after retrieval" technique.) These students would know how many answers to omit on the SAT before they even write it. Widespread strategic preparation of this sort could result in substantial test outcome improvements, particularly for students with a natural tendency to deviate to a large degree from optimal bias.

What other factors affect criterion setting?

The top third of the class had the most accurate "guess" responses, suggesting that they did not compensate perfectly for their higher quality answers by adjusting their report criterion. Had they done so, the accuracy of the "guess" responses would have decreased and possibly matched that in the other two groups. On the other hand, Table 3 shows that students were sensitive to their own level of ability to some extent. That is, averaged across all three tests, the top students guessed at only 0.19 of the questions, the middle third guessed at 0.23, and the bottom third guessed at 0.34. Thus, it appears that participants were sensitive to an important variable—their own level of knowledge—when deciding on how many questions to place in the "guess" category, but that there was definitely room for improvement.

Two other factors that students should have been sensitive to in setting their report criterion were the penalty for errors and level of difficulty of the test. Because Test 1 had a higher penalty and was more difficult than either Test 2 or Test 3, then all other things being equal, students should have placed more responses in the "guess" category when writing it. Table 3 shows that, generally speaking, students were sensitive to these factors in setting their criteria. That is, averaged across rank, actual bias was 0.32 for Test 1, whereas it was only 0.19 for Test 2, and 0.24 for Test 3. Nonetheless, despite sensitivity to these factors, responding remained overly conservative in all cases.

Although responding was generally too conservative regardless of rank and type of test, it is interesting to note the performance of the bottom third of the class on the first test. As shown in Figure 5, these students seemed to be all over the place in terms of their criterion setting. It is possible that these students were overwhelmed by the first test to the extent that it affected their strategic performance. Although their responding stabilised by the second and third tests to resemble the overly conservative performance of the other students, their level of conservatism was greater (i.e., they deviated from optimal bias to a greater degree). This finding suggests a link between student ability and strategic test-taking abilities, such as criterion setting.

CONCLUSIONS

The analyses presented in this paper demonstrated that the corrected score obtained on formula-scored tests is an amalgamation of several different parameters, including metacognitive monitoring and bias. Some researchers and educators may argue that the distinction between knowledge (what you know) and judgements regarding that level of knowledge (what you know about what you know; i.e., monitoring) is an artificial one—both represent the basic concept of knowledge for a particular topic, and the goal of an exam is to measure this overarching concept of knowledge. However, unless these two types of knowledge can be shown to always covary across students and testing situations, binding them together within a single corrected test score can muddy possible interpretations of performance; different combinations of knowledge and metacognitive monitoring can lead to both the same and different outcomes under formula-scoring rules. For instance, in the first exam there were two students who earned identical corrected scores of 61%, but one of the students had almost double the monitoring score of the other ($d' = 1.35$ and 0.78, respectively). Conversely, another set of students on Test 1 had almost identical d' scores of 1.05 and 1.07, yet one of these students scored almost 20% higher than the other student on the corrected score (60% vs. 43%). These two examples highlight the fact that identical test scores do not necessarily reflect identical skills, and that two scores that are different from one another may nonetheless hold aspects in common. Therefore, ignoring metacognitive monitoring by assuming it is a subset of general knowledge (i.e., that they are qualitatively the same) would be a misrepresentation, subsequently leading to a more pallid understanding of what the corrected test score is telling us about an individual's ability.

REFERENCES

Banks, W. P. (1970). Signal detection theory and human memory. *Psychological Bulletin, 74*, 81–99.

Bliss, L. B. (1980). A test of Lord's assumption regarding examinee guessing behavior on multiple-choice tests using elementary school students. *Journal of Educational Measurement, 17*, 147–153.

Clarke, F. R., Birdsall, T. G., & Tanner, W. P. (1959). Two types of ROC curves and definitions of parameters. *Journal of Acoustical Society of America, 31*, 629–630.

Cross, L. H., & Frary, R. B. (1977). An empirical test of Lord's theoretical results regarding formula scoring of multiple-choice tests. *Journal of Educational Measurement, 14*, 313–321.

Ebel, R. L. (1968). Blind guessing on objective achievement tests. *Journal of Educational Measurement, 5*, 321–325.

Galvin, S. J., Podd, J. V., Drga, V., & Whitmore, J. (2003). Type 2 tasks in the theory of signal detectability: Discrimination between correct and incorrect decisions. *Psychonomic Bulletin and Review, 10*, 843–876.

Higham, P. A. (2002). Strong cues are not necessarily weak: Thomson and Tulving (1970) and the encoding specificity principle revisited. *Memory and Cognition, 30*, 67–80.

Higham, P. A. (2007). No special K! A signal detection framework for the strategic regulation of memory accuracy. *Journal of Experimental Psychology: General, 136*, 1–22.

Holzinger, K. J. (1924). On scoring multiple-choice tests. *Journal of Educational Measurement, 15*, 445–447.

Kelemen, W. L., Winningham, R. G., & Weaver, C. A., III. (2007). Repeated testing sessions and scholastic aptitude in college students' metacognitive accuracy. *European Journal of Cognitive Psychology, 19*, 689–717.

Koriat, A., & Goldsmith, M. (1996). Monitoring and control processes in the strategic regulation of memory accuracy. *Psychological Review, 103*, 490–517.

Lord, F. M. (1975). Formula scoring and number-right scoring. *Journal of Educational Measurement, 12*, 7–11.

Muijtjens, A. M. M., van Mameren, H., Hoogenboom, R. J. I., Evers, J. L. H., & van der Vleuten, C. P. M. (1999). The effect of a "don't know" option on test scores: Number-right and formula scoring compared. *Medical Education, 33*, 267–275.

Perfect, T. J., & Stollery, B. T. (1993). Memory and metamemory performance in older adults: One deficit or two? *Quarterly Journal of Experimental Psychology, 46A*, 119–135.

Rawson, K. A., & Dunlosky, J. (2007). Improving students' self-evaluation of learning for key concepts in textbook materials. *European Journal of Cognitive Psychology, 19*, 559–579.

Sax, G., & Collet, L. (1968). The effects of differing instructions and guessing formulas on reliability and validity. *Educational and Psychological Measurement, 28*, 1127–1136.

Sherriffs, A. C., & Boomer, D. S. (1954). Who is penalized by the penalty for guessing? *Journal of Educational Psychology, 45*, 81–90.

Slakter, M. J. (1968a). The penalty for not guessing. *Journal of Educational Measurement, 5*, 41–144.

Slakter, M. J. (1968b). The effect of guessing strategy on objective test scores. *Journal of Educational Measurement, 5*, 217–221.

Thurstone, L. L. (1919). A method for scoring tests. *Psychological Bulletin, 16*, 235–240.

Traub, R. E., Hambleton, R. K., & Singh, B. (1969). Effects of promised reward and threatened penalty on performance of a multiple-choice vocabulary test. *Educational and Psychological Measurement, 29*, 847–861.

APPENDIX

The following is a description of the mathematics used to generate bias profiles. The corrected score can be expressed as a function of the H rate (HR), the FA rate (FAR), the uncorrected raw score (f), and the penalty for errors (p),

$$\text{Corrected score} = HR \cdot f - p \cdot FAR \cdot (1 - f) \tag{1}$$

Now consider the contingency table shown in Table A1. The a, b, c, and d cells in the table represent frequencies, but it is possible to calculate the probabilities of P(a), P(b), P(c), P(d) as follows,

TABLE A1

The 2 × 2 contingency table used to derive the various measures discussed in the Appendix

	Response	
Response category	Correct	Incorrect
Answer	a	b
Guess	c	d

Hit rate (HR) = a/(a + c); false alarm rate (FAR) = b/(b + d); miss rate (MR) = c/(a + c).

$$P(a) = a/(a + b + c + d) = HR \cdot f \tag{2}$$
$$P(b) = b/(a + b + c + d) = FAR \cdot (1 - f) \tag{3}$$
$$P(c) = c/(a + b + c + d) = f - P(a) \tag{4}$$
$$P(d) = d/(a + b + c + d) = (1 - f) - P(b) \tag{5}$$

Now it is possible to express the probability of guesses as function of H, FA, and f.

$$P(\text{guess}) = P(c) + P(d) \tag{6}$$
$$P(\text{guess}) = (f - HR \cdot f) + [(1 - f) - FAR \cdot (1 - f)] \tag{7}$$
$$P(\text{guess}) = 1 - HR \cdot f - FAR \cdot (1 - f) \tag{8}$$

However, because the ROC analysis from Experiment 2 in Higham (in press) indicated that a simple equal-variance Gaussian SDT model accounted for the data, we can express HR as a function of FAR as,

$$HR = \Phi(zFAR + d') \tag{9}$$

where $\Phi(x)$ refers to the probability under the standard normal distribution associated with a z score equal to x, and zFAR refers to the z score corresponding to the FA rate. This term can then be substituted for HR in Equation 1 to yield,

$$\text{Corrected score} = \Phi(\text{zFAR} + d')\cdot\text{f} - \text{p}\cdot\text{FAR}\cdot(1-\text{f}) \qquad (10)$$

The same substitution can be made for the P(guess) in Equation 8 to yield,

$$\text{P(guess)} = 1 - \Phi(\text{zFAR} + d')\cdot\text{f} - \text{FAR}\cdot(1-\text{f}) \qquad (11)$$

Thus, both the corrected score and P(guess) are now expressed as a function of the FA rate, d' (monitoring), the uncorrected, raw score (f), and the penalty for errors (p). It is particularly the critical substitution of $\Phi(\text{zFAR} + d')$ for HR in Equations 10 and 11 that allows a prediction about optimal bias setting to be made. By fixing d', f, and p, and varying FAR from 0 to 1, different values of the corrected score can be obtained using Equation 10, and different values of P(guess) can be generated using Equation 11. It is then possible to plot corrected scores against corresponding P(guess) values—the *bias profile*.

With a slight modification, the bias profile can incorporate more complex payoff matrices than the one shown above. For example, in the current research, the point system not only incorporated a penalty for errors, but also a small reward for misses (c in Table A1). This small reward can be added onto Equation 10 to give,

$$\text{Corrected score}_b = \Phi(\text{zFAR} + d')\cdot\text{f} - \text{p}\cdot\text{FAR}\cdot(1-\text{f}) + \text{b}\cdot\text{MR}\cdot\text{f}$$

$$(12)$$

where Corrected score$_b$ is the corrected score for a test including a bonus for correct guesses (misses), b is the value of that bonus, and MR is the miss rate. To eliminate the MR term, it can be replaced with (1 − HR), which, in turn, can be replaced with $\Phi(\text{zFAR} + d')$. That is,

$$\text{Corrected score} = \Phi(\text{zFAR} + d')\cdot\text{f} - \text{p}\cdot\text{FAR}\cdot(1-\text{f}) + \text{b}\cdot(1 - \text{HR})\cdot\text{f}$$

$$(13)$$

Corrected score

$$= \Phi(\text{zFAR} + d')\cdot\text{f} - \text{p}\cdot\text{FAR}\cdot(1-\text{f}) + \text{b}\cdot[1 - \Phi(\text{zFAR} + d')]\cdot\text{f} \quad (14)$$

Thus, the bias profiles for the research presented in this paper were generated from Equations 11 and 14.

EUROPEAN JOURNAL OF COGNITIVE PSYCHOLOGY
2007, 19 (4/5), 743–768

A cognitive-science based programme to enhance study efficacy in a high and low risk setting

Janet Metcalfe

Department of Psychology, Columbia University, New York, NY, USA

Nate Kornell

Department of Psychology, University of California, Los Angeles, USA

Lisa K. Son

Barnard College, New York, NY, USA

In three experiments, learning performance in a 6- or 7-week cognitive-science based computer-study programme was compared to equal time spent self-studying on paper. The first two experiments were conducted with grade 6 and 7 children in a high risk educational setting, the third with Columbia University undergraduates. The principles the programme implemented included (1) deep, meaningful, elaborative, multimodal processing, (2) transfer-appropriate processing, (3) self-generation and multiple testing of responses, and (4) spaced practice. The programme was also designed to thwart metacognitive illusions that would otherwise lead to inappropriate study patterns. All three experiments showed a distinct advantage in final test performance for the cognitive-science based programme, but this advantage was particularly prominent in the children.

A basic foundation for school accomplishment is the availability of higher order cognitive and metacognitive competencies to realistically assess one's knowledge, to allocate and organise study time and effort optimally, to apply cognitive principles (such as deep, elaborative rehearsal, self-generation, testing, and spacing of practice, to name just a few) that effectively enhance

Correspondence should be addressed to Janet Metcalfe, Department of Psychology, Columbia University, New York, NY 10027, USA. E-mail: jm348@columbia.edu

This research was supported by a grant from the James S. McDonnell Foundation, and by Department of Education CASL Grant No. DE R305H030175. We thank Brady Butterfield, Bridgid Finn, Jason Kruk, Umrao Sethi, and Jasia Pietrzak, and the principal, teachers, and students of MS 143.

Two of the three experiments presented here were described briefly in a short report in the Observer (Metcalfe, 2006). However, the data presented there included only the strict scores, and not the lenient scores, which are presented here.

http://www.psypress.com/ecp   DOI: 10.1080/09541440701326063

learning, and to resist the distractions that could undermine even the most sincerely endorsed intentions. In the population that we targeted in this research, a population of middle school children in an at-risk school setting, these competencies were strikingly limited. Their enhancement was the primary objective of this research. In addition, even among sophisticated learners, specific limitations exist in the use of certain metacognitive strategies. Children and adults often think they know things when they do not (see Metcalfe, 1998; Rawson & Dunlosky, 2007 this issue), and hence inappropriately terminate self-controlled study efforts, or otherwise study in a manner that fails to optimise learning (Bjork, 1994). We sought to devise a computer-assisted study programme, based on principles of cognitive science, that would allow such metacognitive illusions to be overcome.

The project reported here focused on memory enhancing principles derived from experimental studies in cognitive science. Most of these principles, because they have been studied primarily with a focus on understanding the mechanisms underlying memory and cognition, rather than on efforts to facilitate children's academic success, have been investigated in single-session experiments with scholastically irrelevant materials with college-student participants. Despite the apparent lack of connection of this cognitive science literature to real problems that children face in school, to the extent that the principles of cognitive science have some generality, we posited that by implementing them we should be able to enhance learning. The particular cognitive-science principles that we endeavoured to implement included (1) meaningful, elaborative, multimodal processing, (2) test-specific or transfer-appropriate processing, (3) self-generation and multiple testing of responses, and (4) spaced practice. In each of these areas, we focused on overcoming maladaptive metacognitive illusions that might otherwise mislead the student into studying inappropriately. We elaborate on each of these principles below, and review the most relevant research literature.

MEANINGFUL, ELABORATIVE MULTIMODAL PROCESSING

It is now well established that when people process material shallowly—at a perceptual, rote, or nonsemantically informed level—their memory is worse than if they process it deeply, meaningfully, or semantically (Cermak & Craik, 1979; Craik & Lockhart, 1972; Craik & Tulving, 1975), although the explanation for this levels-of-processing effect is still debated (Baddeley, 1978; Metcalfe, 1985, 1997; Nelson, 1977). Memory is critically dependent on people's having a schematic framework providing the deep meaning for the material without which memory performance deteriorates. For example, Bransford and Johnson (1973) showed that (1) when a schema was presented before a story, subjects' comprehension ratings were high and their memory for the story was also high, (2) when the schema was not presented at all subjects' comprehension ratings were low and their memory was low, and (3) when the schema was presented *after* the story, there was a mismatch between the subjects' metacognitive judgements and their performance: Comprehension was rated as high, but memory was low. This apparent conflict between the individual's metacognition and performance may be especially problematic for children, who may feel that once they have understood a set of material, no further study is needed. Hence, in the computer controlled condition we devised a study programme that emphasised the meaningful deep structure of the materials, and provided elaborate rich examples. But to thwart the metacognitive illusion, we also tested the children and reexposed them to the to-be-learned material even in circumstances where they might spontaneously deem it unnecessary.

College students, in a free recall situation, rehearse by interweaving items (Metcalfe & Murdock, 1981; Murdock & Metcalfe, 1978; Rundus, 1971) into meaningful, coherent stories and images. In contrast, Korsakoff amnesics— whose memory performance is seriously deficient—rehearse repetitiously only the last item presented (Cermak, Naus, & Reale, 1976; Cermak & Reale, 1978; and see Metcalfe, 1997)—a strategy that has been shown, even with normal participants, to have little beneficial effect on memory (Bjork, 1970, 1988; Geiselman & Bjork, 1980). Children, who have little experience with studying, may be unaware of the need for integrative and meaningful rehearsal, and may feel that mere rote repetition is enough. Lack of insight into the need for elaborative rehearsal seems especially likely since it has been shown that feelings of knowing can be increased by mere priming (e.g., Reder, 1987; Schwartz & Metcalfe, 1992), and hence this aspect of effective encoding is critical. Moreover, people fail to realise that elaborative rehearsal is more effective than maintenance rehearsal (Shaughnessy, 1981). In the computer programme we devised, to overcome these metacognitive illusions, we mimicked meaningful, elaborative rehearsal by presenting materials in varied contexts, and by using multimodal presentation. We also presented items in several different contexts rather than just one.

TRANSFER APPROPRIATE PROCESSING, OR ENCODING SPECIFICITY

Tulving (1983; Tulving & Thomson, 1973; and more recently, see Roediger, Gallo, & Geraci, 2002) has emphasised that encoding is effective only to the extent that it overlaps with the operations required at the time of retrieval. This encoding specificity principle is inherent in most formal theories of human memory (e.g., Hintzman, 1987; Metcalfe, 1982, 1985, 1995) and has considerable empirical support (e.g., Fisher & Craik, 1977; Hannon & Craik, 2001; Lockhart, Craik, & Jacoby, 1976). Metacognitive research, however, suggests that people are unlikely to take the test situation into account on their own. For example, it is well-established that people make judgements of learning based on heuristics such as how easy they find it to recall the to-be-learned information at the time the judgement is made (Benjamin, Bjork, & Schwartz, 1998; Dunlosky & Nelson, 1994; Dunlosky, Rawson, & McDonald, 2002; Kelley & Lindsay, 1993; Metcalfe, 2002; Weaver & Kelemen, 1997). But if the test conditions do not correspond to the conditions at the time of judgements, the judgements will be inaccurate, and people may, therefore, study inappropriately. For example, when Dunlosky and Nelson (1992) provided participants with targets as well as cues, judgements of learning were inaccurate. Similarly, Jacoby and Kelley (1987) asked participants for their ratings of objective difficulty on a task that was intrinsically very difficult (solving anagrams). Those subjects who were given unsolved anagrams were much more accurate (and thought the problems were more difficult) than people who were given the solutions. These results are important because children, while studying, may routinely look up answers, rather than trying to produce them. Such a strategy will induce illusory confidence. In summary, children may put themselves into the position of believing that they know information when in truth they do not, because they do not test themselves. The study programme implements a self-testing procedure that closely mimics the criterion test itself, and is thereby designed to overcome this illusion.

SELF-GENERATION OF RESPONSES

There is considerable support for the idea that learning is facilitated when people actively attempt to remember and generate responses themselves, rather than passively processing information spoon-fed to them by someone else. In experiments in which college students generated (as compared to read) free associates or rhymes, memory was enhanced (e.g., Schwartz & Metcalfe, 1992; Slamecka & Graf, 1978). The effect has been shown to occur with words (Jacoby, 1978), sentences (Graf, 1980), bigrams (Gardiner & Hampton, 1985), numbers (Gardiner & Rowley, 1984), and pictures (Peynircioglu, 1989), so long as the format of the test is the same as the format of study (Johns & Swanson, 1988; Nairne & Widner, 1987). In most studies, the items have been carefully selected such that in the generate condition, the participant always generated the correct answer. So, for example, in the generate condition, the cue might be Fruit: B_____, and the person had to say banana, whereas in the read condition the stimulus was Fruit: Banana, and the person had only to read the pair. Many studies have shown effects favouring the generation condition. These effects have also been obtained when the answer was not so obvious, and required effortful retrieval from memory, which has been shown to lead to more learning than simply being presented with the answer (Bjork, 1975, 1988; Carrier & Pashler, 1992; Cull, 2000).

Thus far, though, there has been only one study investigating whether students opt to self-generate (Son, 2005). In that study, first graders made judgements of difficulty on cue–target vocabulary pairs. Then they were presented with only the cues and asked if they wanted to read or generate the targets. Results showed that the learners chose to self-generate, particularly when they judged the item to be easy. It remains unknown, however, whether the children would have opted to self-generate had both the cue and target been present when they made the choice. For example, when studying from a textbook, all of the information is present, and therefore, students may be unable to effectively self-generate, insofar as the presence of the answer thwarts efforts to generate it independently. By reading the answer, they have inadvertently put themselves into the "read" condition, which has been shown to be disadvantageous. We used a self-generation procedure in the current study programme.

SPACED PRACTICE

One of the most impressive manipulations, shown repeatedly to enhance learning, is spaced practice. There is now a large literature indicating that, for a wide range of materials, when items are presented repeatedly, they are remembered better if their presentations are spaced apart rather than massed (for a review see Cepeda, Pashler, Vul, Wixted, & Rohrer, 2006). This effect may be of considerable pedagogical importance insofar as it has been shown by Bahrick and Hall (2005; and see Rohrer, Taylor, Pashler, Wixted, & Cepeda, 2005) to hold over extremely long periods of time, up to many years.

An "exception" to the spacing effect has been observed when the test occurs very soon after the last presentation of massed pairs (Glenberg, 1976, 1979). This reversal results in a compelling metacognitive illusion, however, and one that may be exceedingly difficult for the individual to overcome. Because of the immediate efficacy, people trained under massed practice are better pleased with the training procedure and give more favourable judgements about their learning than do those trained under spaced practice (Baddeley & Longman, 1978; Simon & Bjork, 2001; Zechmeister & Shaughnessy, 1980), even though eventual performance favours the latter group. For this reason, even adults, let alone children, do not intuitively understand the benefits of spaced practice, and spacing is unlikely to be used spontaneously; for example, recent data from first graders has shown that children tend to choose massing strategies over spacing strategies (Son, 2005). Therefore, we implemented spaced practice in our learning programme. Two types of spacing were implemented, both of which have been shown to have positive effects: Spacing of items within a study session, and spacing of study sessions themselves.

METACOGNITION AND MOTIVATION

While many well-established findings within the scientific study of human memory show promise of contributing to the effectiveness of the study programme, people's metacognitions are often systematically distorted— mitigating *against* the spontaneous adoption of these effective study strategies. The past 10 years of intensive research has resulted in a growing understanding of the heuristic basis of the mechanisms underlying meta-cognitive judgements (Bjork, 1999; Dunlosky et al., 2002; Koriat, 1994, 1997; Koriat & Goldsmith, 1996; Metcalfe, 1993; Son & Metcalfe, 2005). Metacognitions do not appear to depend upon a person's privileged access to the true state of their future knowledge, but instead, on the information that the person has at hand at the moment the judgement is made. When that information is nondiagnostic or misleading, the judgements will be incorrect, and may lead to inappropriate study.

Deep understanding is both essential for memory and a goal of education. To extract, understand, appreciate, and articulate the core deep meaning of learning material, students need to be able to judge their learning in relation to an ideal state of deep understanding. Because surface knowledge and fluency can influence judgements of knowing, such a result is not automatic. For example, Metcalfe, Schwartz, and Joaquim (1993) showed that repeating a cue paired with the wrong target—a condition that hurt memory performance—resulted in inflated feeling-of-knowing judgements. Reder (1987) and Schwartz and Metcalfe (1992) found that priming parts of a question (e.g., "precious stone") increased subjects' feeling of knowing (e.g., on "What precious stone turns yellow when heated?") without changing the likelihood of the correct answer. Reder and Ritter (1992) found that exposure to parts of an arithmetic problem (e.g., $84 + 63$) inappropriately increased subjects' feeling of knowing to a different problem with surface overlap (e.g., 84×63). Oskamp (1962) provided psychiatrists and psychiatric residents with nondiagnostic information about a patient. The *irrelevant* information increased confidence without changing the accuracy of the diagnosis. The impact of surface knowledge on people's feelings-of-knowing is pernicious, because the inappropriately inflated judgements indicate spuriously to the student that material is understood.

When people are asked how well they will do on a test, their confidence often overshoots actual measured performance (Bandura, 1986; Fischhoff, Slovic, & Lichtenstein, 1977; Lichtenstein & Fischhoff, 1977; Metcalfe, 1986a, 1986b). Students who were about to make an error on an insight problem exhibited especially high confidence (Metcalfe, 1986b), and students who are performing poorly tend to be more overconfident than better students (Fischhoff et al., 1977). This line of research suggests two things: (1) It is important to guard against people's metacognitive feelings, since they are often misleading, and (2) the illusion of overconfidence might be offset, and performance enhanced, if students can be biased to expect that a test will be difficult, rather than easy. The risk in stressing the difficulty of learning materials in realistic educational contexts, however, is that the children may believe that the task is simply impossible, or worse, that they are not smart enough to do it, and hence give up (e.g., Bandura, 1986; Cain & Dweck, 1995; Kamins & Dweck, 1997). This risk may be especially high in the targeted minority population, given Steele's findings on the effects of stereotype threat, a predicament that can handicap members of any group about whom stereotypes exist (e.g., Steele & Aronson, 1995).

In the studies reported here we have tried to create circumstances which would allow students to make realistic judgements about their degree of learning, in several ways. One was by testing them online, as they studied. Another was by telling them at the outset, and continuing to emphasise throughout, that the task itself was extremely difficult. At the same time, we made every effort to bolster motivation, by framing the study as a game— including applause and other entertaining rewards and sounds in the programme, by calling the tutors "coaches", and telling the children that they were playing to score points, and to beat their own previous scores. Indeed, the computer pitted the child against his or her own previous performance, underlining the incremental framework advocated by Dweck (see 1990), rather than ever comparing the child's performance to that of others (which could bolster an entity framework).

THE TARGETED SCHOOL ENVIRONMENT

The experiments reported here are a follow up to those in Metcalfe (2006). Our investigations were especially targeted to a population of inner-city children in a large (1375 students) public middle school, MS 143 in New York City's South Bronx. The children in this school were at potentially high risk for school failure and a wide range of other negative behavioural and social-emotional outcomes. More than one-third of the students had literacy and academic performance scores below New York State's minimal acceptability standards. The study focused on both grades 6 and 7, the approximate age at which highly refined study skills are becoming essential. The study programme, which will be detailed below, attempted to help the children to adopt effective, intrinsically motivated study patterns, informed by effective memory, metacognitive and planning strategies.

Our goal in conducting the experiments described below was to try to maximise the effectiveness of the teaching programme by combining the cognitive principles described above. By doing so, we hoped to demonstrate the potential such a programme has for improving learning in schools and during homework. The design does not allow us to identify the size of the role any particular factor played. We compared the computer study condition to a self study condition in which students were allowed to study as they saw fit. While we did not tell the children what to do in the self study condition, we tried to provide them with every opportunity to implement the motivational, metacognitive, and cognitive principles we implemented in the computer programme, by, for example, providing a calm quiet learning environment, materials with which to make study aids such as flashcards, and all of the words from previous sessions to encourage spaced practice.

EXPERIMENT 1

In the first experiment, we constructed a set of vocabulary items, in consultation with the children's teachers, which were deemed to be important in helping the children to understand the materials that they were studying in class, and useful in allowing them better comprehension of their science and social studies textbooks. These materials were the targets of study.

Method

Participants

The participants in this experiment were 14 children of whom eight sixth and seventh grade children completed all sessions. Six children who started the experiment did not complete it, two because they left the after-school programme and four because they were absent for too many sessions. The participants were students at a poorly performing public middle school, in New York City's South Bronx. They were treated in accordance with the guidelines of the American Psychological Association.

Design

The experiment was a 3 (study condition: computer study, self study, no study) × 2 (test condition: paper, computer) within-participant design. Study condition was manipulated within participants, with vocabulary items assigned randomly, for each child, to one of the three conditions. Test condition was also manipulated within participants; items were tested either on paper or on the computer. This factor was included to test the possibility that there might be an advantage to computer testing in the computer study condition, and an advantage to paper testing in the self study condition. The order of conditions (paper or computer first) was a between-participants factor.

Materials and apparatus

The materials were 131 definition-word pairs; for example, Ancestor–A person from whom one is descended; an organism from which later organisms evolved. Each pair was associated with three sentences, with a blank for the target word; for example, "The mammoth is a(n) _____ of the modern elephant; Venus knew her _____(s) came from Africa and wanted to travel there to explore her roots; Bob was surprised when he found out that his _____(s) were from Norway, because this meant that he was part Norwegian."

The computer portion of the experiment was conducted using Macintosh computers. During self study, in addition to the to-be-learned materials, the children were provided with materials to make study aids, including paper, blank index cards, pens, pencils, and crayons. Self and computer study occurred in a classroom in the children's school as part of an after-school programme.

Procedure

The experiment consisted of seven sessions, which took place once a week. The first session was a pretest. Sessions 2–5 were the main learning phase of the experiment, when all of the words were introduced. Session 6 was a final review of all of the words, and Session 7 was a final test. Children participated in both the computer study and self study conditions within a given session, with the order of conditions balanced across participants.

Pretest Session 1. Participants were given a pretest to identify words they already knew. Any word answered correctly (allowing small spelling errors) was removed from the participant's pool of words, and never shown to that participant again. Spelling errors, here and in our test scoring, were assessed by a computer algorithm based on the amount of overlap between the set of letters used in the subject's answer and the set of letters in the correct answer, and also takes letter order into account. Using this algorithm, the program produced a degree of correctness score from 0 to 100. Small errors were those that had a numerical score between 85 and 99.

During the pretest, the 131 definitions were presented one at a time, in random order. The child's task was to type in the corresponding word and press return. (Although there was some variation in speed, none of the children had problems typing.) Questions remained on the screen for a maximum of 30 s, after which, if the child had not made a response, the question disappeared, and the response was scored as being incorrect. Once a question had either been answered or had timed out, a "Next" button was shown, and when it was pressed, a new question appeared. The entire pretest was conducted on the computer. The pretest continued until the participant got 120 items wrong (which all participants did).

Training (Sessions 2–5).   Sessions lasted for 35 min of self study and 35 min of computer study. During each session, half of the students did computer study and half did self study for 35 min, and then they switched places. Twenty new words were presented during each session, 10 for self study and 10 for computer study.

During self study, the students were given sheets of paper containing each of the 10 new word/definition pairs and their accompanying sentences. The word/definition pair was on one side of the paper, while the three contextual sentences (which included a blank for the target word in question, accompanied by each target word) were on the other side. Students, with their study aids in hand, were allowed to study however they saw fit. At the end of each session all of the study materials the participant had been given or created were put into a folder. At the start of the next session they were given their folder, so that if they chose to, they could study words from all of the preceding sessions, using the materials we had provided and the study aids they had created. Although the children in this condition had the opportunity to space their practice, they were not obligated to do so.

During self study, the experimenters made every effort to ensure that the students paid attention: The classroom environment was quiet and calm, the desks were separated so that the children could not distract each other, and when it seemed to be necessary, the tutors approached the children one-on-one and encouraged them to stay on task. We did this in an effort to equate (or at least increase the equivalence of) motivation between self and computer study, based on the expectation that the computer condition would naturally be more engaging than the self study condition, in part because of the cognitive principles we implemented (e.g., testing), and in part because of the simple fact that it was interactive. It seemed clear that the children behaved during self study in the same way they behaved during class. Because of the small group of students and the presence of multiple tutors, the quality of the self study conditions we provided, when compared to the situations the children normally studied in at home and in school, bordered on being unrealistic.

Computer study during Sessions 2–5 consisted of three main phases. The first phase was a test. During the test, all of the words that had been presented in previous sessions were tested. There was no test in Session 2, because it was the first real session; in Session 3, the 10 words from Session 2 were tested; in Session 4, the 20 words from Sessions 2 and 3 were tested; in Session 5, all 30 words from the previous sessions were tested. During the test at the beginning of each session, the definition and one of the three accompanying sentences were presented, and participants were asked to type in the target word. Each response had to be at least three letters long. Correct responses that were spelled perfectly were followed by a recording of applause played by the computer. If the participant did not respond after 30 seconds the response was automatically coded as incorrect and the computer gave a 'bad' – but slightly funny – beep, followed by the correct word. Responses that were close, but not spelled correctly, were followed by a recording saying, "Close, it's" followed by a recording of the word. The word was then shown on screen. If the response was incorrect a bad beep was played, followed by a recording of the word, and the word was shown on the screen. After the feedback, a "Next" button appeared.

The second phase of computer study was new item presentation. Each of the day's new words was shown individually, along with its definition and an accompanying sentence. Each target/definition/sentence compound was presented for 10 s.

The third phase consumed the majority of the session. In this phase, participants studied the current session's 10 new words. The definition and one of the three accompanying sentences was shown, and the participant was asked to type in the target. The programme cycled through the sentences. If the response was correct, a recording of the word was played, the word was shown on the screen, and the "Next" button appeared. When the "Next" button was pressed, the next definition and accompanying sentence were presented. If the response was incorrect, a bad beep was played followed by a recording of the word, and then the "Next" button appeared. When the "Next" button was pressed, the programme presented the same word again, and continued to do so, using the same procedure, until a correct response was given. From the fourth incorrect guess onwards, the correct answer was shown onscreen at the same time as it was spoken. For close responses that were not spelled correctly, instead of the bad beep, a recording saying, "Close, it's" was played followed by the word, and the word was shown on the screen (regardless of which trial it was).

The definitions were presented repeatedly, in cycles, such that all of the items were first presented in random order, and then, once that cycle was complete, they were all presented again in a rerandomised order. For example, if there had been only four items instead of ten, the order of the first three cycles might have been 2,1,3,4 ... 4,2,3,1 ... 2,1,3,4.

As study progressed, words that were considered temporarily learned were removed from the pool of cycling words to be presented. To be considered temporarily learned a word had to be answered correctly, on the first try, on two separate occasions. In this way, the focus was put on the words that the participant had not learned, and time and effort were not wasted on words they already knew (see Kornell & Metcalfe, 2006; Metcalfe & Kornell, 2005, for more details concerning the efficacy of this strategy).

Study continued until all of the words had been answered correctly twice. At that point, the 10 words that the participant had been studying were added back into the study cycle, along with all of the words that had been answered incorrectly on the pretest, in random order. From this point on, no words were removed from the cycle, and study continued until the 35-min session ended.

At the end of the session, all of the words that the participant had answered correctly on the first try twice, or had answered correctly on the pretest at the start of the session, were displayed on the screen as positive reinforcement, and the participants were told that these were the words they had learned today.

Before beginning the first training session (Session 2), the children were given verbal instructions describing the experiment. They were told that over the course of the next few sessions they would be learning the words that they had seen during the pretest. It was explained that each day they would be given the chance to study words from the previous sessions, as well as new words. They were told that they would have a folder for self study, which would contain all of the materials they had worked on, so that when they started a new session they would be able to use whatever materials they had created in previous sessions. They were also told that the materials they would be learning would be words they would need to know for their classes in school.

Final review (Session 6).   During the final review no new words were presented. For self study, the participants were simply given the folder with the four sheets of words, plus any study aids they had made in the previous weeks. For computer study, the session began with a test on all of the words. Then study commenced on all of the words that were not answered correctly on the test, in the same way as in previous weeks. If and when all of the words had been taken out of the study cycle, all 40 words were added back into the cycle, and study continued with no words being removed, until the 35-min session was complete.

Final test (Session 7).   During the final test all 120 words were tested. Sixty were tested on the computer and sixty were tested on paper, with half of the items from each condition (computer study, self study, no study) being included on each test. On the paper test, the 60 definitions were shown with blank spaces where the words were to be filled in. On the computer test, the definitions were shown one by one, and the participant was asked to type in the response and press return. No feedback was given on either test.

Results and discussion

The data were analysed using an ANOVA with two factors: type of study (computer study, self study, no study) and type of test (computer, paper). The dependent variable was performance on the final test. Performance was scored in two ways, leniently and strictly. With strict scoring, only answers spelled correctly were considered correct, whereas with lenient scoring close answers that were spelled incorrectly were considered correct. The two methods of scoring resulted in the same patterns of significance across all three experiments, so we present only the lenient data. To score the children's responses on the paper test, after the data were collected, the experimenters entered the children's responses on the paper test into a computer program, which scored them, so that the same scoring algorithm was used for both types of test.

The results are shown in Figure 1. There was a significant effect of type of study, $F(2, 14) = 42.19$, $p < .0001$, $\eta_p^2 = .86$ (effect size was computed using

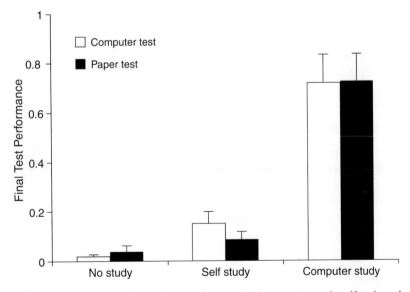

Figure 1.   Experiment 1, children's final test performance in the computer study, self study, and no study conditions. Scores are presented for both the computer test and paper test. Error bars represent standard errors.

partial eta squared). The main effect of type of test was not significant, nor was the interaction. Post hoc Tukey tests showed that performance in the computer study was significantly better than self and no study conditions. The self study condition was not significantly better than the no study condition.

The day on which a word was first studied had a significant effect on final test performance. An analysis of words studied in the computer and self conditions (the no study condition was excluded because the words were not studied on any day) showed that the earlier an item was introduced, the greater its chance of being recalled on the final test, $F(3, 21) = 3.16$, $p < .05$, $\eta_p^2 = .31$. This effect can be explained by the fact that words presented in earlier sessions were restudied in later sessions; as a result, they were presented more times and with greater spacing than words introduced in later sessions.

The results were overwhelmingly favourable for the cognitive-science based programme. Encouraged, we designed Experiment 2 to investigate the efficacy of a similar programme, intended to teach English vocabulary to students who were native Spanish speakers learning English as a second language, in the same at-risk school.

EXPERIMENT 2

Besides the fact that the materials were Spanish–English translations, the main difference between this experiment and Experiment 1 was that the children were actively trying to learn the targeted words in their classrooms. The materials were relatively easy, and were, for the most part, high frequency words referring to common everyday objects.

Method

Participants

The participants were 25 sixth and seventh grade children from the same school as Experiment 1. Seven children who started the experiment but did not finish, four because they left the after-school programme and three because they were absent for too many sessions, leaving eighteen children who completed all sessions. These children were native Spanish speakers, with varying levels of English competence, who were in their school's English as a Second Language (ESL) programme.

Design

The design was a within-participants design with two conditions: computer study and self study.

Materials and apparatus

The materials were 242 Spanish–English translations, e.g., flecha–arrow. In cases in which multiple Spanish synonyms seemed appropriate for an English word, more than one Spanish word was shown, e.g., escritorio, bufete–desk. Pictures of the items were taken from the Snodgrass picture set (Snodgrass & Vanderwart, 1980).

The computer portion of the experiment was conducted using Macintosh computers. During self study, paper, blank index cards, pens, pencils, and crayons were available for the children to make study aids. Self and computer study occurred in a classroom in the children's school as part of an after-school programme.

Procedure

The experiment consisted of six sessions, which took place once a week. The first session was a pretest. Sessions 2–5 were the main learning phase of the experiment, when all of the words were introduced. Session 6 was a final test. The instructions and conditions of self study were similar to Experiment 1.

Pretest Session 1. During the pretest, each of the 242 Spanish words was presented one at a time on the computer. The child's task was to type in the English equivalent and press return. No feedback was given about the accuracy of the responses. Questions could be on the screen for a maximum of 30 s, after which the question disappeared, the response was scored as incorrect, and a button labelled "Listo/a" ("Next") was shown. The reason for the pretest was to identify words the participants already knew. Any word answered correctly (with correct spelling) was taken out of the participant's pool of words, and never shown to that participant again.

Training (Sessions 2–5). Training sessions were divided into two parts, self study and computer study, each of which lasted 30 min. During each session, half of the students engaged in computer study and half in self study, and then they switched places. Forty new words were presented during each session, twenty for self study and twenty for computer study.

During self study, the students were given sheets of paper with each word in both Spanish and English, as well as a picture of the object. They were also given access to blank paper, index cards, pencils, crayons, and pens, so that they could make flashcards or other study aids. As in Experiment 1, participants were allowed to study however they saw fit. At the end of each session all of the study materials the participant had been given or created were put into a folder. At the start of the next session they were given their folder, so that if they chose to, they could study words from all of the preceding sessions, using the materials we had provided and the study aids they had created.

Computer study during Sessions 2–5 consisted of three main phases. The first phase was a pretest (distinct from the session one pretest). During the pretest the "unlearned" translations were presented. The "unlearned" translations included all items that had been presented in previous sessions but had not been answered correctly on subsequent pretests. Being answered correctly on a pretest was the only way for a word to be considered "learned". The pretest did not include the current session's new items; thus, there was no pretest during Session 2, because there were no words that had already been presented. In Session 3, all 20 words from Session 2 were included on the pretest. For example, if the participant answered 7 items correctly on the pretest in Session 3, the remaining 13 words would be pretested again in Session 4 (along with the 20 new words from Session 3).

During the pretest, each Spanish word was presented individually, and participants were asked to type in the English translation. Correct responses that were spelled perfectly were followed by a rewarding "ding" and then a recording of the word being spoken. If the response was close, but not spelled correctly, a recording saying, "Close, it's" was played, followed by a recording of the word, and the word was shown on screen (all of the instructions were in Spanish except for "Close, it's"). If the response was incorrect the recording of the word was played, and the word was shown on the screen. After this feedback, the "Next" button appeared.

The second phase of computer study was initial presentation. Each pair of words was presented once, one pair at a time. First the Spanish word was shown. After 1 s the English word was shown, and then after another 1 s a picture of the item was shown and the word was spoken aloud simultaneously. One second later, the screen cleared and the next word appeared.

The third phase of computer study, during which the session's 20 new words were presented and studied, consumed the majority of the session. The procedure for each presentation during study was as follows: Either the Spanish word or a picture of the item was presented, and the participant was asked to enter the English response. If the response was correct, a beep was played followed by a recording of the word, and the "Next" button appeared. When the "Next" button was pressed, a new word was presented. If the response was incorrect, a bad beep was played followed by a recording of the word, and then the next button appeared. When the "Next" button was pressed, the programme presented the same word again, and continued to do so, using the same procedure, until a correct response was given. From the fourth incorrect guess onwards, the correct answer was shown onscreen at the same time as it was spoken. For close responses that were not spelled correctly, instead of the bad beep, a recording saying, "Close, it's" was played followed by the word, and the word was shown on the screen.

The words were presented repeatedly, in cycles, such that all of the words were presented in random order, and then they were all presented again in a rerandomised order, and so on. For example, if there had been only 4 new items instead of 20, the order of the first three cycles might have been 2,1,3,4 ... 4,2,3,1 ... 2,1,3,4. As study progressed, words that were considered temporarily learned were removed from the cycle of words to be presented. In order to be considered temporarily learned a word had to be answered correctly, on the first try, on two separate occasions. In this way, the focus was put on the words that the participant had not learned. To insure that the participant could answer based on both verbal and pictorial cues, once a correct answer had been given to a picture, the first presentation was always a word from then on, and vice versa.

Study continued until all of the words had been answered correctly twice (or the session ended). At that point, the 20 new words that the participant had just been studying, along with all of the words that had been answered incorrectly on the pretest, were added to the list in random order and study continued. Again, if a word was answered correctly on the first try twice it was removed from the cycle. When all of the words had been removed again, this larger set of words was once again added back into the list, and study continued with none of the words being removed.

At the end of each computer session, the participant was shown two animated cars racing across a track. One represented their memory performance this session, and one their performance in the previous session. This was used as a way to encourage participants to try to do better each session than they had done previously, promoting learning and effort instead of absolute performance.

Final test (Session 6). During the final test 80 self and 80 computer words were presented in random order. Since there was no difference between testing on paper and on the computer in Experiment 1, the entire test was conducted on the computer. The Spanish word was shown, and participants were asked to type in the English word and hit return. Correct answers were followed by a "ding" and then the sound of the word being spoken. Incorrectly spelled answers that were close to correct were followed by a recording that said, "Close, it's" and then a recording of the word. Incorrect answers were followed by a bad beep and then a recording of the word. During the final test, the word was not shown on the screen.

After the 160 words that had been studied were tested, a set of words that had not been studied was tested. The number of unstudied words depended on the number of words the participant had answered correctly on the pretest in Session 1. Because some children had very few (or in one case, zero) unstudied words, we decided to test them at the end of the experiment and did not include them in the data analysis. Test performance for the words that were not studied was lower than for the two other conditions, as expected. The mean accuracy was only 0.14. A score could not be computed for one child who had no words left to be assigned to the no study condition at the end of the pretest.

Results and discussion

As in Experiment 1, there were no differences between strict and lenient scores so only the leniently scored data are presented here. Computer study led to improved performance on the final test, as Figure 2 shows. Performance on the final test was significantly better in the computer study condition ($M = 0.41$) than in the self study condition ($M = 0.29$), $F(1, 17) = 13.61$, $p < .01$, $\eta_p^2 = .44$. Final test performance was also significantly better for items that were first presented in the earlier sessions than items introduced later, $F(3, 51) = 5.08$, $p < .01$, $\eta_p^2 = .23$. As with the first experiment, the results were highly favourable for the cognitively enhanced computer study.

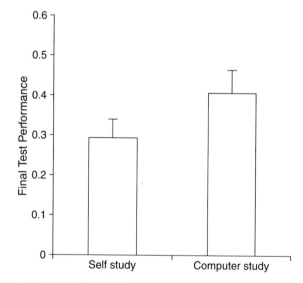

Figure 2. Experiment 2, children's final test performance for the self study and computer study conditions. Error bars represent standard errors.

EXPERIMENT 3

In Experiment 3 we used Columbia University undergraduates as participants. Unlike the young, at-risk children, we expected the undergraduates to have well-developed study skills and to be highly motivated. We expected, though, that even with the highly motivated and skilled Columbia University students, there would still be an advantage for the computer-based study programme, because it circumvented metacognitive illusions, required self generation, and implemented spacing, factors that are difficult to implement without the assistance either of a computer or another person versed in such techniques.

Method

Participants

The participants were 14 Columbia University students who did not speak Spanish. One participant began the experiment but did not complete all of the sessions, leaving 13 complete participants.

Materials and apparatus

The materials were the same 242 Spanish–English translations used in Experiment 2. In Experiment 2, in some cases, multiple Spanish synonyms were used as a single cue. In this experiment, however, the task was to type in the Spanish translation of an English cue. To avoid requiring participants to type in multiple synonyms, we removed all but one of the Spanish synonyms in such cases, for example, changing the pair escritorio, bufete–desk to escritorio–desk. The pictures of the items, which were the same as in Experiment 2, were taken from the Snodgrass pictures (Snodgrass & Vanderwart, 1980).

The computer portion of the experiment was conducted using Macintosh computers. During self study, paper, blank index cards, pens, pencils, and crayons were available for the participants to make study aids. Self and computer study occurred in a laboratory at Columbia University. Like the children in Experiment 2, the students in this experiment were run as groups in a room together.

Design and procedure

The design and procedure of Experiment 3 were the same as Experiment 2, with the following exceptions: In Experiment 3, participants were asked to type in Spanish translations of English words, whereas in Experiment 2 they were asked to type in the English translations of Spanish words. In both experiments, however, the participant's own language was used as the cue and they were asked to type in the language they were learning. Also, the instructions in Experiment 3 were in English instead of Spanish.

Results

The results of Experiment 3, shown in Figure 3, mirrored those of Experiment 2. Performance on the final test was significantly better in the computer study condition ($M = 0.76$) than in the self study condition ($M = 0.67$), $F(1, 12) = 4.96$, $p < .05$, $\eta_p^2 = .29$. Final test performance was also better for words first presented in earlier sessions than words introduced later, $F(3, 36) = 52.52$, $p < .0001$, $\eta_p^2 = .81$. Test performance for words that were not studied was 0.10.

CONCLUSION

In three experiments, we tested the effectiveness of a computer-based learning programme that used principles of cognitive science, including elaborative processing, generation, transfer-appropriate processing, spacing, and metacognitive and motivational techniques. All three experiments showed that the computer-based programme led to significantly more learning than self study. The boost in performance was present for at-risk

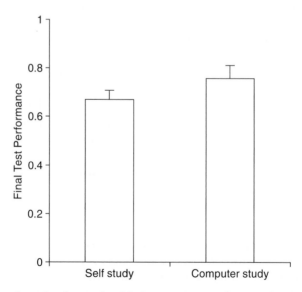

Figure 3.   Experiment 3, college students' final test performance for the self study and computer study conditions. Error bars represent standard errors.

middle school students (Experiments 1 and 2) and for the more experienced college students (Experiment 3).

Our research here was based on the work of previous researchers who have discovered a number of strategies (e.g. generation, testing, and spacing) that have benefited learning in the laboratory. The application of these strategies in the classroom has lagged behind the laboratory research. However, there has been a recent surge in interest in both isolating principles of learning that are likely to have beneficial effects in classroom settings, and in actually implementing such principles to improve children's learning. The efficacy of the translation of some of these components is beginning to emerge. For example, in a series of recent papers, (Chan, McDermott, & Roediger, in press; Kuo & Hirshman, 1996; McDaniel, Anderson, Derbish, & Morrisette, 2007 this issue; Roediger & Karpicke, 2006; Roediger, McDaniel, & McDermott, 2006) it has been shown that not only does testing improve memory more than does additional studying, but that this "testing effect" occurs in the classroom as well as in the lab. We used multiple tests in our programmes, and so it is reassuring to see that this was well justified: When the "testing effect" that we made use of in our composite study has been isolated in classroom situations, it does, indeed, improve learning.

The generation effect, long established in the laboratory, and aggressively used in the computer-based study programmes that we employed here, has a more chequered recent history in classroom situations. Metcalfe and Kornell (in press), and deWinstanley's group (deWinstanley & Bjork, 2004; deWinstanley, Bjork, & Bjork, 1996) have found that the generation effect is not always in evidence when performance on the generate items is compared to a control condition in which items were presented to the students. The reason for this apparent failure of a well-replicated laboratory finding may not be that the generation effect itself is not robust, however, but rather that it may be too robust. It may be very easy to get students, in a classroom or study situation, to mentally generate, even when they are nominally in the control condition. They may simply generate spontaneously or they may generate because they have learned that it is advantageous to their performance. When they do not generate, however, as Metcalfe and Kornell's (in press) follow-up experiments show, performance is impaired both in the lab and in the classroom setting. Thus, evidence is accumulating that the strategy that we used, of forcing the children to generate, was, indeed a good one. One might wonder whether and to what extent the effects of generating are mitigated by the fact that the children sometimes generate errors. In the programmes reported here, many errors were, indeed, generated, though corrective feedback was given.

Finally, while the children were free to space their practice in the self study condition, they were forced to do so in the computer controlled programme, and presumably spacing was used more rigorously in the latter than the former condition in the studies presented here. While the laboratory evidence is overwhelming that spacing is beneficial in this kind of learning, the data from the classroom for the advantages of spacing are only now beginning to emerge. Interestingly, while the memory effects for factual material are holding up, Pashler, Rohrer, and Cepeda (2006) have recently reported that in certain practical domains—such as in the acquisition of visual-spatial skills, or in certain categorisation tasks—spacing is unimportant (but see Kornell & Bjork, in press, for evidence that spacing is effective in a categorisation task). Future research, directed at improving teaching in particular domains and for particular targeted skills and knowledge bases, may reveal important qualifications concerning the generality of spacing effects. The tasks that we explored in our study were classic verbal acquisitions tasks, and as such we expected spacing to be important. However, by taking aim at the materials and concepts that children need to learn in school situations, we may reveal important limitations to what have, almost universally, been accepted as fundamental principals of learning such as the advantage of spaced practice.

The primary goal of this computer-based study was to take effective cognitive and metacognitive strategies out of the laboratory, and implement them in a real learning situation, and to compare the outcome with students' natural learning strategies. The findings show that young learners are often prone to spontaneously using *ineffective* study strategies, and that even accomplished college students, who have a great deal of practice studying effectively, would benefit from direct cognitive and metacognitive learning programmes. The current programme is only a small step towards bridging the gap between laboratory testing and individual learning, but these data show that laboratory research on learning principles can be valuable in enhancing learning in classroom situations. We conclude that learning programmes that employ motivational, cognitive, and metacognitive techniques together are effective learning tools. In the future, it would be of interest to further investigate the magnitude of the role each strategy played in improving learning. Such a research strategy may serve not only to consolidate our confidence about the procedures that we have used, when those procedures survive translation into a classroom setting and are appropriate, but it may also, serendipitously, reveal that some of those supposedly well-established principles are not what they seem, despite a great deal of laboratory research.

REFERENCES

Baddeley, A. D. (1978). The trouble with levels: A re-examination of Craik and Lockhart's framework for memory research. *Psychological Review, 85*, 139–152.

Baddeley, A. D., & Longman, D. J. (1978). The influence of length and frequency of training session on the rate of learning to type. *Ergonomics, 21*, 627–635.

Bahrick, H. P., & Hall, L. K. (2005). The importance of retrieval failures to long-term retention: A metacognitive explanation of the spacing effect. *Journal of Memory and Language, 52*(4), 566–577.

Bandura, A. (1986). *Social foundations of thought and action: A social cognitive theory.* Englewood Cliffs, NJ: Prentice Hall.

Benjamin, A. S., Bjork, R. A., & Schwartz, B. L. (1998). The mismeasure of memory: When retrieval fluency is misleading as a metamnemonic index. *Journal of Experimental Psychology: General, 127,* 55–68.

Bjork, R. A. (1970). Repetition and rehearsal mechanisms in models of short-term memory. In D. A. Norman (Ed.), *Models of memory* (pp. 307–330). New York: Academic Press.

Bjork, R. A. (1975). Short-term storage: The ordered output of a central processor. In F. Restle, R. M. Shiffrin, N. J. Castellan, H. R. Lindman, & D. B. Pisoni (Eds.), *Cognitive theory* (pp. 151–171). Hillsdale, NJ: Lawrence Erlbaum Associates, Inc.

Bjork, R. A. (1988). Retrieval practice. In M. Gruneberg & P. E. Morris (Eds.), *Practical aspects of memory: Current research and issues, Vol. 1: Memory in everyday life* (pp. 396–401). New York: Wiley.

Bjork, R. A. (1994). Memory and metamemory considerations in the training of human beings. In J. Metcalfe & A. J. Shimamura (Eds.), *Metacognition: Knowing about knowing* (pp. 185–205). Cambridge, MA: MIT Press.

Bjork, R. A. (1999). Assessing our own competence: Heuristics and illusions. In D. Gopher & A. Koriat (Eds.), *Attention and performance XVII: Cognitive regulation of performance: Interaction of theory and application* (pp. 435–459). Cambridge, MA: MIT Press.

Bransford, J. D., & Johnson, M. K. (1973). Considerations of some problems of comprehension. In W. Chase (Ed.), *Visual information processing* (pp. 383–438). New York: Academic Press.

Cain, K. M., & Dweck, C. S. (1995). The relation between motivational patterns and achievement cognitions through the elementary school years. *Merrill-Palmer Quarterly, 41,* 25–52.

Carrier, M., & Pashler, H. (1992). The influence of retrieval on retention. *Memory & Cognition, 20,* 633–642.

Cepeda, N. J., Pashler, H., Vul, E., Wixted, J. T., & Rohrer, D. (2006). Distributed practice in verbal recall tasks: A review and quantitative synthesis. *Psychological Bulletin, 132,* 354–380.

Cermak, L. S., & Craik, F. I. M. (1979). *Levels of processing in human memory.* Hillsdale, NJ: Lawrence Erlbaum Associates, Inc.

Cermak, L. S., Naus, M. J., & Reale, L. (1976). Rehearsal strategies of alcoholic Korsakoff patients. *Brain and Language, 3,* 375–385.

Cermak, L. S., & Reale, L. (1978). Depth of processing and retention of words by alcoholic Korsakoff patients. *Journal of Experimental Psychology: Human Learning and Memory, 4,* 165–174.

Chan, J. C. K., McDermott, K. B., & Roediger, H. (in press). Retrieval-induced facilitation: Initially nontested material can benefit from prior testing of related material. *Journal of Experimental Psychology: General.*

Craik, F. I., & Lockhart, R. S. (1972). Levels of processing: A framework for memory research. *Journal of Verbal Learning and Verbal Behavior, 11,* 671–684.

Craik, F. I. M., & Tulving, E. (1975). Depth of processing and the retention of words in episodic memory. *Journal of Experimental Psychology: General, 104,* 268–294.

Cull, W. L. (2000). Untangling the benefits of multiple study opportunities and repeated testing for cued recall. *Applied Cognitive Psychology, 14,* 215–235.

DeWinstanley, P. A., & Bjork, E. L. (2004). Processing strategies and the generation effect: Implications for making a better reader. *Memory & Cognition, 32,* 945–955.

DeWinstanley, P. A., Bjork, E. L., & Bjork, R. A. (1996). Generation effects and the lack thereof: The role of transfer-appropriate processing. *Memory, 4,* 31–48.

Dunlosky, J., & Nelson, T. O. (1992). Importance of the kind of cue for judgments of learning (JOL) and the delayed-JOL effect. *Memory & Cognition, 20*, 374–380.

Dunlosky, J., & Nelson, T. O. (1994). Does the sensitivity of judgments of learning (JOLs) to the effects of various study activities depend on when the JOLs occur? *Journal of Memory and Language, 33*, 545–565.

Dunlosky, J., Rawson, K. A., & McDonald, S. L. (2002). Influence of practice tests on the accuracy of predicting memory performance for paired associates, sentences, and text material. In T. Perfect & B. Schwartz (Eds.), *Applied metacognition* (pp. 68–92). Cambridge, UK: Cambridge University Press.

Dweck, C. S. (1990). Self-theories and goals: Their role in motivation, personality, and development. In R. A. Dienstbier (Ed.), *Nebraska symposium on motivation* (Vol. 38, pp. 199–235). Lincoln, NE: University of Nebraska Press.

Fischhoff, B., Slovic, P., & Lichtenstein, S. (1977). Knowing with certainty: The appropriateness of extreme confidence. *Journal of Experimental Psychology: Human Perception and Performance, 3*, 552–564.

Fisher, R. P., & Craik, F. I. M. (1977). Interaction between encoding and retrieval operations in cued recall. *Journal of Experimental Psychology: Human Learning and Memory, 3*, 701–711.

Gardiner, J. M., & Hampton, J. A. (1985). Semantic memory and the generation effect: Some tests of the lexical activation hypothesis. *Journal of Experimental Psychology: Learning, Memory, and Cognition, 11*, 732–741.

Gardiner, J. M., & Rowley, J. M. (1984). A generation effect with numbers rather than words. *Memory & Cognition, 12*, 443–445.

Geiselman, R. E., & Bjork, R. A. (1980). Primary and secondary rehearsal in imagined voices: Differential effects on recognition. *Cognitive Psychology, 12*, 188–205.

Glenberg, A. M. (1976). Monotonic and nonmonotonic lag effects in paired-associate and recognition memory paradigms. *Journal of Verbal Learning and Verbal Behavior, 15*, 1–16.

Glenberg, A. M. (1979). Component levels theory of the effects of spacing of repetitions on recall and recognition. *Memory & Cognition, 7*, 95–112.

Graf, P. (1980). Two consequences of generating: Increased inter- and intraword organization of sentences. *Journal of Verbal Learning and Verbal Behavior, 19*, 316–327.

Hannon, B., & Craik, F. I. M. (2001). Encoding specificity revisited: The role of semantics. *Canadian Journal of Experimental Psychology, 55*, 231–243.

Hintzman, D. L. (1987). Recognition and recall in MINERVA2: Analysis of the "recognition-failure" paradigm. In P. Morris (Ed.), *Modeling cognition* (pp. 215–229). New York: Wiley.

Jacoby, L. L. (1978). On interpreting the effects of repetition: Solving a problem versus remembering a solution. *Journal of Verbal Learning and Verbal Behavior, 17*, 649–667.

Jacoby, L. L., & Kelley, C. M. (1987). Unconscious influences of memory for a prior event. *Personality and Social Psychology Bulletin, 13*, 314–336.

Johns, E. E., & Swanson, L. G. (1988). The generation effect with nonwords. *Journal of Experimental Psychology: Learning, Memory, and Cognition, 14*, 180–190.

Kamins, M. L., & Dweck, C. S. (1997, May). *The effect of praise on children's coping with setbacks.* Poster presented at the ninth annual meeting of the American Psychological Society, Washington, DC.

Kelley, C. M., & Lindsay, D. S. (1993). Remembering mistaken for knowing: Ease of retrieval as a basis for confidence in answers to general knowledge questions. *Journal of Memory and Language, 32*, 1–24.

Koriat, A. (1994). Memory's knowledge of its own knowledge: The accessibility account of the feeling of knowing. In J. Metcalfe & A. P. Shimamura (Eds.), *Metacognition: Knowing about knowing* (pp. 115–135). Cambridge, MA: MIT Press.

Koriat, A. (1997). Monitoring one's own knowledge during study: A cue-utilization approach to judgments of learning. *Journal of Experimental Psychology: General, 126*, 349–370.

Koriat, A., & Goldsmith, M. (1996). Monitoring and control processes in the strategic regulation of memory accuracy. *Psychological Review, 103*, 490–517.

Kornell, N., & Bjork, R. A. (in press). The promise and perils of self-regulated learning. *Psychonomic Bulletin and Review.*

Kornell, N., & Metcalfe, J. (2006). Study efficacy and the region of proximal learning framework. *Journal of Experimental Psychology: Learning, Memory, and Cognition, 32*, 609–622.

Kuo, T., & Hirshman, E. (1996). Investigations of the testing effect. *American Journal of Psychology, 109*, 451–464.

Lichtenstein, S., & Fischhoff, B. (1977). Do those who know more also know more about how much they know? *Organizational Behavior and Human Performance, 20*, 159–183.

Lockhart, R. S., Craik, F. I. M., & Jacoby, L. L. (1976). Depth of processing, recognition, and recall: Some aspects of a general memory system. In J. Brown (Ed.), *Recall and recognition* (pp. 75–102). London: Wiley.

McDaniel, M. A., Anderson, J. L., Derbish, M. H., & Morrisette, M. (2007). Testing the testing effect in the classroom. *European Journal of Cognitive Psychology, 19*, 494–513.

Metcalfe, J. (1982). A composite holographic associative recall model. *Psychological Review, 89*, 627–661.

Metcalfe, J. (1985). Levels of processing, encoding specificity, elaboration, and CHARM. *Psychological Review, 92*, 1–38.

Metcalfe, J. (1986a). Feeling of knowing in memory and problem solving. *Journal of Experimental Psychology: Learning, Memory, and Cognition, 12*, 288–294.

Metcalfe, J. (1986b). Premonitions of insight predict impending error. *Journal of Experimental Psychology: Learning, Memory, and Cognition, 12*, 623–634.

Metcalfe, J. (1993). Novelty monitoring, metacognition, and control in a composite holographic associative recall model: Implications for Korsakoff amnesia. *Psychological Review, 100*, 3–22.

Metcalfe, J. (1995). Distortions in human memory. In M. A. Arbib (Ed.), *The handbook of brain theory and neural networks* (pp. 321–322). Cambridge, MA: MIT Press.

Metcalfe, J. (1997). Predicting syndromes of amnesia from a Composite Holographic Associative Recall/Recognition Model (CHARM). *Memory, 5*, 233–253.

Metcalfe, J. (1998). Cognitive optimism: Self-deception or memory-based processing heuristics? *Personality and Social Psychology Review, 2*, 100–110.

Metcalfe, J. (2002). Is study time allocated selectively to a region of proximal learning? *Journal of Experimental Psychology: General, 131*, 349–363.

Metcalfe, J. (2006). Principles of cognitive science in education. *APS Observer, 19*, 27–38.

Metcalfe, J., & Kornell, N. (2005). A region of proximal learning model of study time allocation. *Journal of Memory and Language, 52*, 463–477.

Metcalfe, J., & Kornell, N. (in press). Principles of cognitive science in education: The effects of generation, errors and feedback. *Psychonomic Bulletin and Review.*

Metcalfe, J., & Murdock, B. B., Jr. (1981). An encoding and retrieval model of single-trial free recall. *Journal of Verbal Learning and Verbal Behavior, 20*, 161–189.

Metcalfe, J., Schwartz, B. L., & Joaquim, S. G. (1993). The cue-familiarity heuristic in metacognition. *Journal of Experimental Psychology: Learning, Memory, and Cognition, 19*, 851–861.

Murdock, B. B., Jr., & Metcalfe, J. (1978). Controlled rehearsal in single-trial free recall. *Journal of Verbal Learning and Verbal Behavior, 17*, 309–327.

Nairne, J. S., & Widner, R. L. (1987). Generation effects with nonwords: The role of test appropriateness. *Journal of Experimental Psychology: Learning, Memory, and Cognition, 13*, 164–171.

Nelson, T. O. (1977). Repetition and depth of processing. *Journal of Verbal Learning and Verbal Behavior, 16*, 151–172.

Oskamp, S. (1962). The relationship of clinical experience and training methods to several criteria of clinical prediction. *Psychological Monographs, 76*, 28.

Pashler, H., Rohrer, D., & Cepeda, N. J. (2006). Temporal spacing and learning. *Observer, 19*, 19.

Peynircioglu, Z. F. (1989). The generation effect with pictures and nonsense figures. *Acta Psychologica, 70*, 153–160.

Rawson, K. A., & Dunlosky, J. (2007). Improving students' self-evaluation of learning for key concepts in textbook materials. *European Journal of Cognitive Psychology, 19*, 559–579.

Reder, L. M. (1987). Strategy selection in question answering. *Cognitive Psychology, 19*, 90–138.

Reder, L. M., & Ritter, F. E. (1992). What determines initial feeling of knowing? Familiarity with question terms, not with the answer. *Journal of Experimental Psychology: Learning, Memory, and Cognition, 18*, 435–451.

Roediger, H. L., Gallo, D. A., & Geraci, L. (2002). Processing approaches to cognition: The impetus from the levels of processing framework. *Memory, 10*, 319–322.

Roediger, H. L., & Karpicke, J. D. (2006). Test-enhanced learning: Taking memory tests improves long-term retention. *Psychological Science, 17*, 249–255.

Roediger, H. L., McDaniel, M. A., & McDermott, K. B. (2006). Test-enhanced learning. *APS Observer, 19*, 28–38.

Rohrer, D., Taylor, K., Pashler, H., Wixted, J. T., & Cepeda, N. J. (2005). The effect of overlearning on long-term retention. *Applied Cognitive Psychology, 19*, 361–374.

Rundus, D. (1971). Analysis of rehearsal processes in free recall. *Journal of Experimental Psychology, 89*, 63–77.

Schwartz, B. L., & Metcalfe, J. (1992). Cue familiarity but not target retrievability enhances feeling-of-knowing judgments. *Journal of Experimental Psychology: Learning, Memory, and Cognition, 18*, 1074–1083.

Shaughnessy, J. J. (1981). Memory monitoring accuracy and modification of rehearsal strategies. *Journal of Verbal Learning and Verbal Behavior, 20*, 216–230.

Simon, D. A., & Bjork, R. A. (2001). Metacognition in motor learning. *Journal of Experimental Psychology: Learning, Memory, and Cognition, 27*, 907–912.

Slamecka, N. J., & Graf, P. (1978). The generation effect: Delineation of a phenomenon. *Journal of Experimental Psychology: Human Learning and Memory, 4*, 592–604.

Snodgrass, J. G., & Vanderwart, M. (1980). A standardized set of 260 pictures: Norms for name agreement, image agreement, familiarity, and visual complexity. *Journal of Experimental Psychology: Human Learning and Memory, 6*, 174–215.

Son, L. K. (2005). Metacognitive control: Children's short-term versus long-term study strategies. *Journal of General Psychology, 132*, 347–363.

Son, L. K., & Metcalfe, J. (2005). Judgments of learning: Evidence for a two-stage model. *Memory and Cognition, 33*, 1116–1129.

Steele, C. M., & Aronson, J. (1995). Stereotype threat and the intellectual test performance of African Americans. *Journal of Personality and Social Psychology, 69*, 797–811.

Tulving, E. (1983). *Elements of episodic memory.* Oxford, UK: Oxford University Press.

Tulving, E., & Thomson, D. M. (1973). Encoding specificity and retrieval processes in episodic memory. *Psychological Review, 80*, 352–373.

Weaver, C. A., III, & Kelemen, W. L. (1997). Judgments of learning at delays: Shifts in response patterns or increased metamemory accuracy? *Psychological Science, 8*, 318–321.

Zechmeister, E. B., & Shaughnessy, J. J. (1980). When you know that you know and when you think that you know but you don't. *Bulletin of the Psychonomic Society, 15*, 41–44.

Subject index